HYDROLOGY AND WATER RESOURCES IN TROPICAL REGIONS

DEVELOPMENTS IN WATER SCIENCE, 18

OTHER TITLES IN THIS SERIES

1 G. BUGLIARELLO AND F. GUNTER
COMPUTER SYSTEMS AND WATER RESOURCES

2 H. L. GOLTERMAN
PHYSIOLOGICAL LIMNOLOGY

3 Y. Y. HAIMES, W. A. HALL AND H. T. FREEDMAN
MULTIOBJECTIVE OPTIMIZATION IN WATER RESOURCES SYSTEMS.
THE SURROGATE WORTH TRADE-OFF-METHOD

4 J. J. FRIED
GROUNDWATER POLLUTION

5 N. RAJARATNAM
TURBULENT JETS

6 D. STEPHENSON
PIPELINE DESIGN FOR WATER ENGINEERS

7 V. HÁLEK AND J. ŠVEC
GROUNDWATER HYDRAULICS

8 J. BALEK
HYDROLOGY AND WATER RESOURCES IN TROPICAL AFRICA

9 T. A. McMAHON AND R. G. MEIN
RESERVOIR CAPACITY AND YIELD

10 G. KOVÁCS
SEEPAGE HYDRAULICS

11 W. H. GRAF AND C. H. MORTIMER (EDITORS)
HYDRODYNAMICS OF LAKES: PROCEEDINGS OF A SYMPOSIUM
12–13 OCTOBER, 1978, LAUSANNE, SWITZERLAND

12 W. BACK AND D. A. STEPHENSON (EDITORS)
CONTEMPORARY HYDROGEOLOGY: THE GEORGE BURKE MAXEY
MEMORIAL VOLUME

13 M. A. MARIÑO AND J. N. LUTHIN
SEEPAGE AND GROUNDWATER

14 D. STEPHENSON
STORMWATER HYDROLOGY AND DRAINAGE

15 D. STEPHENSON
PIPELINE DESIGN FOR WATER ENGINEERS
(completely revised edition of Vol. 6 in the series)

16 W. BACK AND R. LÉTOLLE (EDITORS)
SYMPOSIUM ON GEOCHEMISTRY OF GROUNDWATER

17 A. H. EL-SHAARAWI AND S. R. ESTERBY (EDITORS)
TIME SERIES METHODS IN HYDROSCIENCES

HYDROLOGY AND WATER RESOURCES IN TROPICAL REGIONS

by

JAROSLAV BALEK

Senior Researcher, Stavební Geologie, Prague, Czechoslovakia

ELSEVIER

Amsterdam — Oxford — New York — 1983

Published in co-edition with
SNTL Publishers of Technical Literature, Prague

Distribution of this book is being handled by the following publishers

for the U.S.A. and Canada
Elsevier Science Publishing Co. Inc.
52, Vanderbilt Avenue
New York, N.Y. 10017

for the East European Countries, China, Northern Korea, Cuba, Vietnam and Mongolia

SNTL, Publishers of Technical Literature, Prague

for all remaining areas
Elsevier Science Publishers, B. V.
Molenwerf 1
P.O. Box 211, 1000 AE Amsterdam, The Netherlands

Library of Congress Cataloging in Publication Data

Balek, Jaroslav.
Hydrology and water resources in tropical regions.

(Developments in water science ; 18)
Includes bibliographies and index.
1. Hydrology--Tropics. I. Title. II. Series.
GB840.B34 1983 553.7'0913 83-5664
ISBN 0-444-99656-7

ISBN 0—444—99656—7 (vol. 18)

ISBN 0—444—41699—2 (Series)

Printed in Czechoslovakia

FOREWORD

The tropical regions of the world represent a huge but relatively little studied part of the globe. The preponderance of less developed countries in the tropical zone means that scientists, teachers and planners often look to the lst or 2nd world for assistance, but within the field of hydrology there are relatively few comprehensive tomes. Individual papers and huge collections of disparate conference presentations tend to predominate but this text tries to encompass the whole of the tropics and covers the field of water from condensation in the atmosphere to its use and disposal by man.

We are never allowed to forget that the Tropical World is different (from the temperate zones). The 15 000 mm of annual rainfall at Debundja in Cameroun would take over 25 years to fall in London whilst a fish production of 200 kg ha^{-1} in 120 days in paddy fields in Malagasy is hardly credible by the standards of British fisheries. The Moslem principle that water is common property is shown to guide water resources development in many countries in a direction quite different to those areas following the Roman Law principles of private, common and public ownership. Moreover, we are reminded that in some tropical areas sources of surface water, though harmless to the local populace, may produce serious diseases in outsiders.

This is not a standard hydrology book nor is it an engineering manual. Its subject is hydrology and the possibilities and problems of using water resources. It is about the complex system involving hydrology, ecology and human activity which some authors describe as a geographical approach to hydrology. Throughout the orientation is a wide ranging systems approach. Human impact on hydrology is usually seen as negative but the onward effects nutrient cycling are discussed. Floods are described not solely as a hazard to be eliminated but as long awaited events in dry areas.The Aswan dam is the subject of a social audit as well as a hydrological review. Modest scale irrigation schemes are advocated with the proviso that the planning team needs a health expert included. Finally, the impact of other mammals is discussed with examples cited of erosion induced by elephants and lake levels influenced by hippos grazing swamp vegetation.

Jerry Balek is particularly able to take this broad but authoritative view of water schemes. His rich experience of work from Indonesia to Albania, India to Zambia and Uganda plus his command of a multilingual bibliography is woven into the text. His own studies permeate each chapter and it is no coincidence that one of the hydrological models discussed is called DAMBO after the African word for swamp. The section on the exploration phase of hydrology reflects another of Jerry's interests for emphasis is given to the exploration of the River Niger by a team using canoes.

Dr. G. E. Hollis
Department of Geography,
University College London.

PREFACE

Several years ago I published a book on hydrology and water resources in tropical Africa [1]. At that time the number of references relating to hydrological problems of the African tropics was very limited. Since then, however, an increasing interest in the water economy of the developing countries, the majority of which are located within the boundaries of the tropics, has resulted in the publication of a great many papers, reports and texts on tropical waters and their management.

This book is concerned with the behaviour of tropical waters and various ecological, geographical and climatological conditions. The problems of water management in relation to agriculture and civil engineering are also examined.

Tropical regions today are faced with extensive and rapid changes. Unfortunately man's impact on tropical hydro-ecology is predominantly negative. Coordinated action to prevent the situation from deteriorating further should be based on a thorough understanding of natural water regimes and the changes and development they undergo. If this book can at least contribute in some small degree to a better knowledge of tropical waters, then the work devoted to it has not been in vain.

Readers of this book will find the list of references to be more extensive than in the previous one. And one can well believe that many valuable interim reports still remain hidden in libraries and archives. Knowing of these works relating to a wide variety of tropical basins could lead to the further development of tropical hydrology. I would thus be very grateful to recieve any publication in the field of tropical water resources that the reader may know of, even if it has so far enjoyed only local or limited circulation.

J. Balek

[1] Balek, J., 1977. Hydrology and water resources in tropical Africa, Elsevier, Amsterdam, 208p.

CONTENTS

1 THE WORLD OF TROPICAL RIVERS

Anthony Smith (*The Seasons*):
"...With considerable myopia the people who live in the temperate parts of the world tend to assume that their kind of season applies all over the planet, that the days shorten as winter approaches, the leaves fall, that the nights grow chilly, then grow cold, then freezing, then longer, then warmer, and hot once again while the new leaves bask in the summer sun. Temperate people should spare a thought for the tropical regions where seasons are more a matter of rainfall, or deluge followed by droughts, and where the wind may blow consistently from one direction for half the year, only to blow with equal consistency for the second half. Seasons, in short, are no one thing, but they are all different manifestations of that most singular property of the planet Earth, namely that it revolves around the sun with its axes at an angle to the plane of its rotation."

1.1 INTRODUCTION

Through the ages people have been compelled to settle in regions where water was deficient in quantity or erratic in supply. When supplies failed or were made useless or when floods swept everything away, then and only then were centres of habitation abandoned. Man's endeavours to achieve a better relationship with the waters of the earth have helped to mould his character and his outlook towards the world around him.

It was along tropical rivers that the earliest stages of man's biological and cultural development originated. Pithecantropus erectus lived on Java 400 – 500 thousand years ago and in the Upper Pleistocene a man related to him who is believed to have lived 35 – 75 thousand years ago, was found in the deposits of the river Solo. Most recent remnants of modern Homo sapiens dating back 40 000 and 20 000 years have been found in Borneo and the Philippines respectively. A few hundred ape-man fossils believed to be several million years old have been found in East and South Africa. Fossils of an ancestor of man found in 1974 in the Awash Valley, Ethiopia are believed to be 3 – 4 million years old. Deposits at Olduvai, Tanzania produced Zinjanthroupus and Homo habilis, a creature using the first tools made from bones and pebbles also originated in East Africa one million years ago.

The fluctuation of wet and dry periods has always had a great impact on man's existence in the tropics. Tropical islands such as Borneo, Java and Sumatra were part of the Asian continent about twenty thousand years ago when the ocean level was much lower. Other islands in that region are the results of tectonic activity or are of volcanic origin and the fauna, the flora and also the hydrological regimes reflect a different origin. For instance, a land bridge which once existed between India and Sri Lanka was in the region of tropical rainfall and thus migration existed between the two parts. That is why animals typical of the tropical rainforest moved into Sri Lanka.

The fluctuation of wet and dry periods, the so-called pluvials and interpluvials influenced the earliest stages of man and man's ancestor in the tropics. The pluvials correspond to the glacials periods in the moderate regions during the last million years. The interpluvials or dry periods lasted twenty thousand to two hundred thousand years. During the glacial periods in the temperate regions the mean temperature of the tropics was about 0.8 °C lower than today. In Africa, Wayland [21] identified the main periods of increased rainfall, separated by periods of drier climate, at Lake Victoria. Nilsson [16] found two wet post-pleistocene periods called Makalian and Makuran, another two wet periods in the Late Pleistocene called Gamblian and Nakuran, one wet period in the Early Pleistocene called Kageran. Leakey [12] identified rich cultural history and faunal development during the pluvials.

The regions of equal summer and winter rainfall probably remained the same throughout the Quaternary, but small changes resulted in the shifting of ecological zones. These shifts were conceivably extensive. For instance, during Late Acheulian and Aterian times the Sahara Desert was invaded by Mediterranean flora down to its southern limits. More temperate conditions and an annual rainfall in excess of 600 mm existed.

In recent history the impact of the fluctuation of dry and wet periods can be traced more easily. Historians have proved that the resettlement of the Mayas occurred as a result of changing hydro-ecological conditions. The Maya empire flourished between the second and eighth centuries A.D. and the original cities were abandoned at the beginning of the tenth century A.D. The foundation of a new empire south and north of the former site followed and lasted until the sixteenth century, and it was already in state of decay when the Spaniards finally destroyed it. The migration can be explained by climatic changes, however, some historians claim that under conditions of shifting cultivation the increasing demands of a developed society made necessary a move away from the exhausted soil of the original site.

Other authors support a theory that the Mayan culture of Yucatan was developed through an extensive use of groundwater; later the original wells had to be abandoned because of serious problems of pollution resulting from the use of a sewage disposal well adjacent to each supply well.

In fact, many abandoned cities existed on the plains of northern Guatemala which today consists of almost an uninhabited jungle.

The religion of the Mayas was water oriented and priests prayed to Chac, the water god, for his assistance in water management i.e. to decrease the severity of droughts. It is believed that the T-shaped eyes full of tears of the god Chac symbolised rain. Caves and caverns were believed to contain virgin water for religious purposes and the preparation of offerings to Chac. A renowned cenote at Chichen Itza is elliptical with the water level several metres below the rim and a diameter of about 60 metres. It was used as a "sacrificial well" to propitiate the rain god. Some authors believe that the increasing emphasis on human sacrifice reflected the increasing problem of population growth.

An archaelogical survey in the extremely arid region of Chile, the Pampa del Tamarugal, which experiences 1 mm of rain per year, indicated that there were settlements there less than 2000 years ago and that some of them still existed at the time of the colonization by the Spaniards.

Variable hydrologic conditions had a dominant effect on the migration of the human population. As can be traced from hunting scenes on the walls of caves in the Sahara and other parts of Africa, at some time the whole continent was populated by hunters. Later the hunters survived in the lush grasslands and tropical rainforest, herders remained nomadic in their search for pastures and cultivators became tied by sources of water.

In many places, however, man has always preferred to meet his water problems head-on rather than quit his place of abode, agriculture and industry. People started to apply their creative imagination and skills to build the more sophisticated water constructions which are so significant for human advancement.

The Mayas, for instance, practised engineering to control water. Irigoyen [11] described how underground water tanks called chultuns were constructed to collect surface runoff during the rainy periods. A typical chultun could store and provide enough water to support twenty five persons. Also small aquaducts and canals were constructed to carry the water from the hills to areas of settlement. Seasonal supplements of water was provided by aguadas, a type of natural well or a shallow depression lined with stones or impermeable clay.

The Egyptians made the best possible use of Nile observations which enabled them to predict rich and poor crops, and regulate taxes etc. The Nile floods even furnished a reason for complicated religious rites and theological doctrines. Verner [20] concluded that the computation of the hypothetical fluctuation of the Nile's water volume together with present knowledge of Egyptian history, may result in a possible correlation between periods of high-water years and periods of political, economical and cultural development.

Numerous nilometres were constructed on the Nile and some of them have supplied significant data for recent research (Anděl, Balek, Verner [1]).

A sophisticated system of irrigated agriculture still practised by the Kelabites in Central Borneo is operated according to the migration of birds, this being influenced by the monsoonal regime.

The evaluation of water resources, supply storage and distribution has been carried out since ancient times, often with remarkable sophistication. In arid regions deep wells were dug and underground drains driven to considerable depths and over long distances, toward the foothills of the mountains. Extensive aquaducts were constructed to convey spring water, even if often on a trial and error basis.

The habits of man and the forms of his social organisations were influenced more by their close association with water than with the land, resulting in the first water laws. Cervera and Arias [4] analysing the occurrence of droughts in Mexico found that at one time it was prohibited to take corn out of the Valley of Mexico "on pain of death" in years of drought. As reported from the period of drought between 1450 and 1454 A.D., poverty became so great that a person could be sold for the price of corn.

Property rights were associated primarily with the uses of water. Mohammed declared that free access to water was the right of every Muslim community. However, the question of how much water a person living in an arid region should be entitled to is disputable, for example there is a great difference between the needs of Kalahari bushman who can survive on 1 litre per day and an inhabitant of a tropical industrial city with a consumption of more than 800 litres per person per day.

It is clear that the history of mankind is closely related to water resources and water scarcity. The role of rivers and water resources in the history of tropical regions is perhaps even more important.

1.2 THE LATIN AMERICAN TROPICS

The great geographical myth of El Dorado, based on verbal testimony of a lake in the South American interior containing a treasury of gold, led many expeditions between 1533 and the present day to unexplored regions of the Latin American continent. It was in 1535 that Belalcazar with a group of mercenaries explored the country north of Quito. However, the region had already been searched by an expedition under the command of Hernan Perez de Quesada (1500 – 1579) and by another expedition led by N. Federmann (1500 – 1542) who reached the same region from the Venezuelan coast. The ultimate goal of all the expeditions was the mountain lake Guatavitá. The depth of this lake was a hundred meters and it was not possible to reach the bottom. However, the legend survived and led other expeditions to different parts of the continent. Diego de Ordas in 1531 – 1532 searched for the El Dorado in the vicinity of the river Meta in Colombia and on the altiplano of Bolivia. Gonzalo Pizzaro (1511 – 1548) searched the Amazon headwaters. Part of the expedition returned to Peru and part of it commanded by Pizarro's deputy F. Orellana (1505–1546) explored the banks of the Amazon. In 1558 another expedition led by Pedro de Ursúa started for the rivers Huallaga and Marañon. After the murder of Ursúa a new commander, Aguirre, 'General of Marañon' led the expedition. Probably they followed the river Negro and reached the Caribbean coast by way of Orinoco. While Orellana's descent of the Amazon river is known, the exact route of Ursúas-Aguirre's expedition is uncertain. Ramsay [17] believed that many historical maps made by the Spaniards were strictly classified and are still not accessible to this day.

In 1584, Antonio de Berreo became convinced that southeast of the Orinoco there was a lake and a city called Manoa where there was an abundance of gold. An expedition organized by Berreo and led by Domingo de Vera was unsuccessful, nevertheless, the story encouraged Sir Walter Raleigh himself to sail up the Orinoco in an expedition led by L. Keyimis. However, nothing significant was found.

In 1913 when Lake Guatavitá was emptied, a few golden objects were found. Lake Manoa remained only on the map in Raleigh's unpublished manuscript of 1596. In 1802 Alexander von Humboldt explored the Orinoco basin and proved that there were no significant lakes in the region.

Hydrographically more interesting was the search for the connection between the basin of the Amazon and the Orinoco. The existence of the Casiquiare Canal connecting the upper Orinoco with the Rio Negro was first reported by Father Acuña, a Jesuit priest, in 1639 and confirmed before 1744 by another Jesuit, Father Roman. Before Diego de Ordaz explored the Orinoco, Columbus noticed the freshness of the water in the Gulf of Paria in 1498, however, no one attempted to investigate the source.

The first scientific account of a trip to the Orinoco in 1800 was given by Humbold and Bourland in 1807 [10].

Another great river in Latin America, the Paraná was explored as early as 1526–1529 by Sebastian Cabot and from 1542 the first Jesuit missions were established there. Similarly, the Guyana Highlands became accessible by boat, and canoes have plied the Guyanan rivers for four centuries.

The rivers also formed the boundaries between Guyana and Surinam and their sources became a matter of dispute, as in the case of the headwaters of the River Courantyne forming the boundaries between British Guyana and Surinam.

The most significant part of the hydrographical explorations in South America has been concerned with the Amazon basin. The river was surveyed 80 kilometres inland by Vincenta Yañez Pinzon in 1500 and from the very beginning it bore several different names. First it was called Rio Santa Maria de la Mar Dulce, later only Mar Dulce, also Rio Grande and El Rio Marañon. Orellana probably gave it the name Amazon in belief that the women of Tapuyas tribes fought together with their men against the invaders. He accomplished the first descent of the river from the Andes to the ocean, reaching the main stream through the Napo River in 1541.

First ascent of the river by Orellano's route was accomplished by Pedro Texeira in 1638, who followed the River Napo and reached Quito. Other tributaries were explored much later. In 1884 – 1887 the Xingú was explored by Karl von Steinen. In 1910 – 1927 an expedition led by Roosevelt and Rondon canoed along the River Doubt and found it to be a branch of the Madeira River. A survey of the Rivers Negro, Uaupes and Branco was accomplished by Rice between 1910 – 1924.

In 1923 – 1924 the American Rubber Mission supported by the U.S. Government explored thirty-seven rivers including Tocatins, Xingú, Tapajós, Madeira, Mamoré, Beni, Madre de Dios, Aare, Purú, Ucayali, Huallaga, Negro and Branco and not only were the rivers explored but also the soil and agricultural potential along the river banks.

Early in 1925 Colonel P. W. Fawcett and his expedition left Cuyabá to explore the Xingú-Tapajós region and were lost. In 1928 G. M. Dyott of England traced the route of the expedition and found evidence that the expedition had perished.

Shortly before the World War II, H. Rittlinger accomplished a one man descent of the Huallaga-Amazon in a collapsible kayak.

The search for the ultimate sources of the river was accomplished only recently. Formerly the Amazon was regarded as a separate river from the confluence of the Rivers Marañon and Ucayali. In 1953 Perton surveyed the river and claimed that its headwaters should be sought in upper Apurimac. Thus the formerly recognized length of the Amazon increased to 6640 kilometres (Netopil [15]). In 1955 Perton changed his original conclusion and decided that the ultimate source of the river was Lake Kilafro near the Lake Titicaca. In 1958 Cordiche found another source to be the River Huyaco in the Marañon basin. In 1960 in Revista Brasiliera de Geografia, information was published to the effect that the Amazon's ultimate source was the river system of Ucayali-Urubamba and that the length of the River Amazon amounted to 6577 kilometres. Only two years later the Bolivian Geographical Journal claimed that the very ultimate source was the River Apurimac, in its headwaters called Rio Santiago, which originates from the glacier Uacra. Then the length of the river reached 7025 kilometres making it the longest river of the world.

1.3 AUSTRALIA

The history of Australia is perhaps more closely related to its rivers and water resources than the history of any other continent, because most of the explorers searched first of all for surface and subsurface water resources.

From the hydrological point of view the visit to the continent by William Dampier in 1688 should be mentioned because he landed several times in search of water, but described the land as being barren. In 1669 the East India Company organized an expedition which entered the inland and discovered the Swan River. Small expeditions organized from Bathurst resulted in the discovery of the River Lachlan and in 1813 G. Blaxland, W. Lawson and W. C. Wentworth crossed the River Nepean at Emu Plains and discovered the grassy valley of Fish River.

In 1816 Lieutenant Oxley accompanied by Evans and Cunnigham traced the River Lachlan to find that it ended in swamps. Hume and Howell started their expedition from Lake George and via the Murrumbidgee river they reached the River Murray. In 1824 Governor Sir Ralph Darling wanted to trace the routes of rivers flowing westward and organized an expedition in 1828, which was commanded by C. Sturt. On the way from Macquaire marshes they found a large river which they named Darling. In 1831 Sturt followed the Murrumbidgee River, which he described as rapid and deep and followed it for more than 3000 kilometres. He found that both rivers joined the River Murray.

In 1844 Dr. Leichhard on his overland route from Port Victoria to Morelon Bay reached the source of the River Burdekin. However, he perished on another expedition, started in 1844 on the River Condamine in Queensland. The northern tributaries of the river Darling were explored by T. Mitchell and E. B. Kennedy. In 1858 Babbage and Warburton proved that the Lakes Torreno, Eyre and Gairdner are not a single lake as Eyre had originally thought.

Burke and Will's expedition of 1861 directed from Melbourne to the River Flinders at the Gulf of Carpentaria, ended in the death of all but one member of the expedition. This was the fate of many more parties exploring the water resources of the Australian interior. In this case the rescue party under the command of Landsborough explored the Rivers Barcoo, Flinders and Gilbert.

In 1874 J. Forrest who was requested by the Perth Government to explore the basins of the rivers flowing towards the northern and western shores of Australia did not find any streams. Also the expedition of E. Giles which crossed the central Australian desert found the interior to be sandy and waterless.

Other expeditions set up for geologic explorations contributed to knowledge of the hydrological divides and thus the hydrography of the continent was already well known at the beginning of the twentieth century.

1.4 ASIAN RIVERS

In contrast to the tropical rivers of other continents Asian rivers have historically been under the impact of headwaters located north of the tropics. The Brahmaputra, the Indus and the Karnali were said to encircle the sacred area of Mount Kalais seven times before flowing in various directions, thus paying homage to the throne of the gods, according to the ancient rite of circumambulation. The Tibetians considered the Brahmaputra to be a river flowing from the mouths of animals which are the vehicles of the throne symbols of the Dhyani-Budhas (Govinda [8]).

It is through these rivers that religions and cultural relations between Tibet and countries of southern Asia were established and the banks of these rivers are still regarded as sacred places.

In fact not all the four rivers have their sources on the slopes of Mount Kailas, or have their mouths in the geographical tropics, however, they are greatly influenced by the tropical climate. Their valleys lead toward the plateau which is dominated by Mount Kailas, called Meru or Sumeru according to the oldest sanskrit tradition. The mountain is regarded as the physical and metaphysical centre of the world, and because our psycho-physical organism is a microcosmi replica of the universe, Meru is represented by the spinal cord in the nervous system.

Other oriental river basins too have been under the influence of Tibetian culture. The Mekong was followed for a considerable part as a caravan route connecting China with Tibet and another significant caravan route – the Tali-Bhamo route – crossed it. By the Irrawaddy and the Salween, in the remote past descendants of Mongolian tribe migrated from China and Tibet and became the ancestors of present inhabitants of Burma.

The first contacts with southern Asia were mainly commercial. They can be traced back to the fifteenth century when Portuguese, Spanish, Dutch and English ships went in search of spices and other oriental products and landed in India, Ceylon, the Malaysian Peninsula and in the Philippines. A hydrographical survey was initiated much later. The great mapping of India was accomplished in 1843 by Sir George Everest and then the sources of Ganga and Sutley were located.

The Salween basin was explored by J. Crawford in 1830 – 1837. He also visited the middle part of the Mekong while travelling in Thailand and Laos. The first trip from the Irrawaddy to southern Vietnam was made by A. Bastian in 1861 – 1863. A. Pavie explored the Mekong basin in 1886 – 1891 while in 1887 Rimmer examined the navigability of the Irrawaddy. He found it to be navigable 240 kilometres above Bhamo. Gautier made a descent of the river from Luangprabang to its mouth in 1887 – 1888.

Jawa was explored and mapped at the beginning of nineteenth century by W. Junghuhn. R. Wallace travelled widely in the Malaysian archipelago after 1854. The interior of the Celebes was visited between 1893 – 1903 by F. and P. Sarasins.

Not until 1913 were the gorges of Brahmaputra surveyed by Bailey and Morshead who together with Kingdom in 1924 found the connection between the river Tsangpu in Tibet and the Brahmaputra.

1.5 AFRICAN RIVERS

The oldest known hydrological records are from the African continent and the history of the management of water resources in Africa is among the oldest in the world. One of the very earliest records is a drawing of an imperial macehead held by the protodynastic King Scorpion taking part in the ceremony of cutting an irrigation ditch. His reign is dated at about 3200 B.C.

The relatively narrow strip of inundated and therefore cultivated soil on both banks of the Nile has always been bounded by the desert. Since the end of prehistoric times an increase in agriculture can be noted in the inundated zone and it made a surprising leap at the beginning of the Old Kingdom, about 2800 B.C.

Life in the whole country was always dependent on the seemingly uncontrolled whims of the greatest river of Africa. Sufficient water meant rich crops while a shortage of water resulted in poor crops and famine. The effort to unify the disorganized system of water regulation and to build irrigation tracks was a very important, perhaps decisive factor in the emergence and growth of state power. It was the prolonged and thorough observation of the river that enabled the Egyptians to predict rich or poor crops and regulate taxes.

The history of the search for the sources of the river was remarkable. According to Ptolemy, in second century A. D. Diogenes reported two large interior lakes in tropical Africa and a range of snow-capped mountains in the region where the Nile rose, called the Mountains of the Moon. However, it was not until the nineteenth century that the attention of European geographers became focused on the Nile. In 1856 they were still discussing whether the Nile originated from the famous fountains as described by Herodotus.

In 1855 Captain R. F. Burton accompanied by J. H. Speke discovered the Lakes Nayasa, Tanganyika and Victoria. Unlike Speke, Burton was not convinced that Lake Victoria was the source of the Nile. Speke was entrusted by The Royal Geographical Society to prove this and together with J. A. Grant he visited the Nile Falls. In 1866 Livingstone, still unconvinced that the river's true sources had been found, speculated that Lake Nyasa might drain into Lake Tanganyika which might be linked with Lake Albert and with the Nile. He believed that the source of the Congo and the Nile could be identical. In 1872, one year after he was found by Stanley, he died at Bangweulu swamps, still searching for the fountains.

The actual source of the river was not known until 1937 when B. Walddecker traced the southernmost tributary of the Kagera river flowing into the Lake Victoria.

No earlier than in 1923 the Blue Nile canyon was traversed by an expedition led by R. E. Cheesman.

The middle Niger has been an important centre for the west African market since the beginning of the Christian era. The basin reached its zenith in the fourteenth century when it became a real centre for trade between Tunis and Egypt on the one hand and the west coast on the other.

Originally the river was believed to be identical with the Congo. Leo Africanus suggested that the river flowed from an interior source into the ocean in a westerly direction. In 1788 J. Ledyard was entrusted with the task of travelling along the Niger, however, he died before being able to complete his mission.

Mungo Park was entrusted by the Association for the Promoting the Discovery of the Interior of Africa to ascertain the course and the termination of the river. In 1796 he reached the Niger but was able to follow it for only six days.

H. Clapperton who travelled to Lake Chad, accompanied by Dr. W. Oudney and Debenham, visited the Bight of Benin and tried to prove that the Oil Rivers at Brass are identical with the Niger. He died at Sokoto in 1827. His former servant R. L. Lander with his brother John sailed the whole river in two small canoes in 1830.

In 1854 H. Barth navigated the headwaters of the River Benue while his friend A. Overweg made a survey of Lake Chad. In 1873 L. E. Dupont and O. Lenz discovered the River Ogooué.

In 1843 Diego Cao, sailing along the coast of West Africa, discovered a fresh water stream thirty kilometres from the coast and became the first European to land on the banks of the Congo. In 1912 J. K. Tuckey was sent to explore the river, however, the expedition failed completely. In 1866 Livingstone travelled through the Congo headwaters, assuming that Lake Mweru contributed to the Nile's waters.

In 1877 H. M. Stanley arrived at Nyangwe after a 999-day trip. Only 115 men, women and children out of 350 reached the final point. Later it was proved by Thompson that Lake Tanganyika, through the Lukuga outlet, was also part of the Congo system.

The river Kasai, a tributary of Congo, was put on the map by Wissman and the southern tributaries were mapped by Grenfell.

In 1891 P. Crampel was killed when exploring the Sangha and the Chari.

The routes of traders from the Indian Ocean followed another mighty African stream, the Zambezi. Once the African plateau was reached through the escarpments there was easy access to the mining areas of iron, copper and gold and to ivory. A further increase in the importance of the Zambezi River came with the extension of commercial contacts in the fifteenth century. In 1514–1516 A. Fernandes followed the Sabi and Zambezi looking for new routes to the interior. Already in 1700 the Zambezi was plotted as the Cuana or Zambere on the map of F. Morden.

In 1798–99 Dr. L. Lacerda trying to establish a trans-African route along the Zambezi deviated from the Zambezi at Tete to Lake Mweru and to the court of Kazembe; he died there in 1799. Captain Gamitto explored the Luangwa valley in 1831 and in 1854 J. da Silva reached the centre of Lozi Kingdom on the upper Zambezi at Lealui.

Livingstone visited the Victoria Falls on Zambezi in 1855. After reaching Chobe he followed the Zambezi north and then turned in the direction of Luanda. On the way back he canoed the river from Sesheke to the small islands above the falls.

In 1886 Dr. E. Holub operated north of the flood plains of the Zambezi tributary, the River Kafue and there carried out the first hydrographical measurements. Twenty years later, the first hydrographical post was opened nearby at the Kafue Bridge connecting South Africa with the Central African interior. This meant that a new road was opened for scientific hydrology in the tropics.

1.6 REFERENCES

[1] Anděl, J., Balek, J., Verner, M., 1971. An analysis of the historical sequences of the Nile maxima and minima. Symp. on the role of hydrol. in the Econ. Dev. of Africa. WMO Rep. No. 301, Addis Abbaba, 24—36.

[2] Balek, J., 1977. Hydrology and water resources in tropical Africa. Elsevier Amsterdam, 208 p.

[3] Biswas, A. K., 1970. History of hydrology. Elsevier, Amsterdam.

[4] Cervera, J. J., Arias, D. P., 1981. A perspective study of droughts in Mexico. Journal of Hydrology, 51, 4—155

[5] Clark, J. D., 1963. Ecology and culture in the African Pleistocene. S. Afr. J. of Science IX, No. 7, 363—366.

[6] Condamine de la Maria, 1941. Relación abreviada de un viaje hecho por el interior de la Américe Meridional. Espasa-Calpe, S.A., Madrid, 229 p. (First published in 1745).

[7] Freise, F. W., 1938. The drought region of northeastern Brazil. Geog. Review, Vol. 18, 363—378.

[8] Govinda, A., 1970. The ways of the white clouds. Shambhala Boulder, 305 p.

[9] Holub, E., 1890. Von der Kapstadt ins Land der Maschukulumbe. Hölder, Wien.

[10] Humboldt, A., Bourland, A., 1807. The voyage aux régions équinoxiales fait en 1799—1804.

[11] Irigoyen, R., 1970. Bajo el signo de Chac. Monografía del agus potable en Yucatan. ZAMNA. Merida, 135 p.

[12] Leakey, L. S. B., 1964. Prehistoric man in the tropical environment. Symp. on the ecol. of man in the trop. environ., Nairobi. Mogens, 24—29.

[13] Lévi-Strauss, C., 1955. Tristes tropiques. Librairie Plon, Paris, 295 p.

[14] Medina, J. T., 1934. The discovery of the Amazon. Amer. Soc. Gogr. Spec. Publ., No. 17, N. York, 467 p.

[15] Netopil, R., 1972. Hydrology of continents. Academia Prague (In Czech), 296 p.

[16] Nilsson, E., 1932. Quaternary glaciations and pluvial lakes in British East Africa. Geogr. Ann., Stockholm, Vol. 13, 249—349.

[17] Ramsay, R. H., 1972. No longer on the map. The Viking Press, N. York.

[18] Sánchez, Labrador, J., 1910. El Paraguay Católico. Buenos Aires.

[19] Steinen von den, K., 1894. Durch Zentral Brasilien. Leipzig.

[20] Verner, M., 1972. Periodical water volume of the Nile. Arch. Orient., Prague, 105—123.

[21] Wayland, E. J., 1934. Rifts, rivers, rain and early man in Uganda. Lond. R. Ant. Inst. Jour., Vol. 64, 333—352.

[22] Wells, H. G., 1965. A short history of the world. Penguin B. Ltd., Harmondsworth, 364 p.

2 CLIMATOLOGY OF THE TROPICS

2.1 THE TROPICS

The tropical regions are geographically located between the parallels 23°27′ south and 23°27′ north of the equator, referred to as the Tropic of Capricorn and the Tropic of the Cancer, respectively. From a meteorological point of view, the boundaries of the tropics are sometimes taken to be the dividing lines between easterlies and westerlies and thus the region is also sometimes considered to be bound by the latitudes 30° north and south of the equator. However, such a definition based on the most frequently occurring differencies in atmospheric conditions, is only an approximate one because these phenomena are not notably confined to particular latitudes.

The line between the eastern and western winds and the region of the subtropical high pressure cells is not seasonally or geographically stable. If any definition of the tropical regions based on the boundaries of atmospheric circulation is to be accurate, then the kinematic and thermodynamic features of the atmosphere must be more clearly defined.

The definition of the equator is not entirely rigid either. Besides the accurate definition of the geographical equator, a 'meteorological' equator is occasionally refered to; this is considered to be the mean latitude of the equatorial trough at a latitude of 5 °N. Some meteorologists take (perhaps more accurately) the fluctuating line of the Intertropical Convergence Zone (ITCZ) as a meteorological equator.

Likewise the term equatorial heat is used to refer to the line connecting points of highest annual temperature around the earth. Along much of its length, this line runs parallel to the geographical equator, but ranges from 20 °N in Mexico to 14° S in Brazil. The term thermal equator is similar, referring to the line of highest annual temperature which is presumably located close to the latitude 5° north.

One of the main features of tropical regions is the very high absorption of heat and its low emanation. Thus the tropical regions can be seen as a substantial reservoir of heat energy, the shortage of heat at higher latitudes being compensated for by the excess heat in the tropics.

Between Cancer and Capricorn the greatest angle of elevation of the sun during the day ranges between 43° and 90° and daylight extends from 10.75 to 13.25 hours. The maximum possible hours of sunshine for various latitudes in different months are given in Table 2.1.

The sun is directly overhead twice a year and the temperature is under the direct

Tab. 2.1 Maximum possible hours of sunshine at selected latitudes

30 S	20 S	10 S	0	10 N	20 N	30 N	Month
13.9	13.2	12.6	12.1	11.6	11.1	10.4	J
13.2	12.8	12.5	12.1	11.8	11.5	11.1	F
12.4	12.3	12.2	12.1	12.1	12.0	12.0	M
11.5	11.7	11.9	12.1	12.3	12.6	12.9	A
10.7	11.2	11.7	12.1	12.6	13.1	13.6	M
10.3	10.9	11.6	12.1	12.7	13.3	14.4	J
10.4	11.1	11.6	12.1	12.6	13.2	13.9	J
11.1	11.5	11.8	12.1	12.5	12.8	13.2	A
12.0	12.0	12.1	12.1	12.2	12.3	12.4	S
12.9	12.6	12.3	12.1	11.9	11.7	11.5	O
13.6	13.1	12.6	12.1	11.7	11.2	10.7	N
14.0	13.3	12.0	12.1	11.6	10.9	10.3	D

influence of the sun. Water circulation, or the hydrological cycle, is driven by solar energy; one can assume that in the tropics heat from the sun influences the hydrological cycle more strongly than in other regions. However, the relationship between solar energy and the movement of water is far from simple, because the influx of solar energy is not uniform over the earth's surface, but rather evokes a complex system of circulation. At the equator, the sun is overhead on March 21 and September 22 and the influx of solar radiation is greatest at these times. At the solstices, the intensity of the sun's radiation is at a minimum, although the difference between minimum and maximum energy influxes is much lower in the tropics than in the temperate regions.

The solar radiation falling on a hypothetical horizontal surface at the top of the atmosphere and at the earth's surface is shown in Fig. 2.1. The normal fluctuation of solar activity greatly influences the hydrological circulation system, but the relationship implied in this is not easily demonstrable by standard methods of investigations. For instance, the periodicity of various hydrological phenomena is affected not only by solar activity, but also by other factors such as the actual number of hours of sunshine, which may be large in deserts but otherwise tends to be low in the tropics owing to extensive cloud cover. During the hot season the clouds are of the cumulus type produced by convection currents. Continuous sheets of rainclouds occur in the vicinity of low atmospheric pressure, under the influence of air currents from the sea and also local mountain formations may cause the clouds to thicken. Occasionally, fine weather can be interrupted by so-called 'gutti' spells, which are periods of overcast, drizzly weather, also frequently associated with an invasion of sea air currents.

Tab. 2.2 Thermal balance of the surface of the earth and of the atmosphere (10^{-3} Joule cm^{-2} $year^{-1}$). R is the radiation balance of the surface, L is the radiation lost due to evaporation, T is thermal turbulent flow into the atmosphere, F is thermal transport by ocean currents, R′ is the radiation balance of the atmosphere, L′ is the thermal balance of atmospheric condensation

Latitude	Land			Ocean				Land and ocean				Atmosphere	
	R	L	T	R	L	T	F	R	L	T	F	R′	L′
20—30° N	289	80	209	465	444	29	—8	398	306	96	—4	—377	163
10—20° N	297	134	163	515	490	29	—4	456	393	63	0	—381	167
0—10° N	301	239	63	527	444	29	54	477	398	38	41	—352	536
0—10° S	301	255	46	535	406	25	104	481	368	29	84	—339	423
10—20° S	306	188	117	515	473	38	4	465	410	55	0	—381	310
20—30° S	293	117	176	460	452	46	—38	423	377	75	—29	—352	331
Tropics	298	169	129	513	451	33	0	450	375	59	16	—364	322
Earth	205	113	92	381	343	38	0	331	281	50	0	—331	280

If we regard the hydrological regime as a part of the environment, then any study of the variation of the regime with time has to take into account not only the hydrology of river basins but also the regime of the whole hydrosphere including troposphere and perhaps even higher layers which mediate between the sun and the basin surface. The predominance of the solar energy influx as the most significant factor affecting the hydrological regime is increased in climatically homogeneous areas. In small basins there are other factors of greater or lesser importance compared with the solar factor, and local factors frequently play a decisive role.

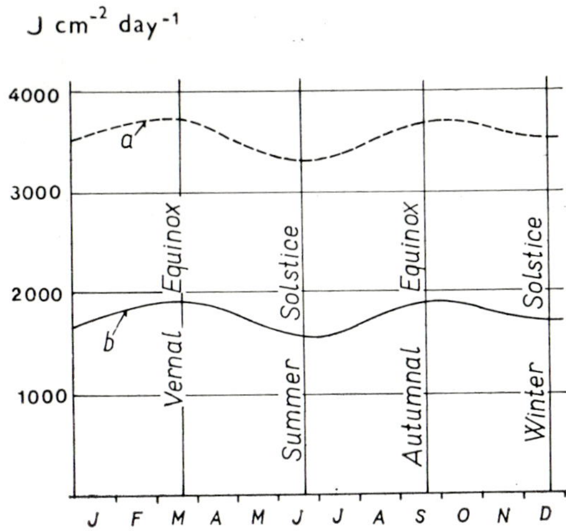

Fig. 2.1 Fluctuation of solar radiation at the equator, a, above the limits of the atmosphere, b, received at the earth's surface

A thermal budgeting drawn up for tropical regions by the Soviet Central Geophysical Obervatory is presented in Table 2.2, while the values of solar radiation at various latitudes of the tropics are in Table 2.3.

As can be seen from the budgeting, about 90% of the energy input to the tropical oceans is absorbed in the process of evaporation, 7% heats the atmosphere, and 3% is carried by ocean currents to higher latitudes. On land, the proportions are different; about 58% of the energy is consumed in evaporation and 42% goes into the atmosphere. The following meteorological features are common to low latitudes:

a) the horizontal variability of meteorological phenomena is small;

b) the periodicity of recurring meteorological events is high;

c) meteorological processes occurring on a moderate scale more often tend to develop into large scale processes.

d) high air moisture of the air in the tropics means that the water plays a more important role in thermal budgeting.

Violent weather conditions are perhaps more common in the tropics than elsewhere. In November 1970 a cyclone along the coast of Bangladesh took a toll

Tab. 2.3 The values of solar radiation in (Joule cm^{-2} day^{-1}) at various latitudes

30 S	20 S	10 S	0	10 N	20 N	30 N	Month
4253	4139	3918	3596	3173	2670	2005	J
3893	3947	3889	3713	3437	3052	2592	F
3345	3592	3730	3746	3663	3449	3152	M
2687	3098	3424	3638	3751	3755	3642	A
2131	2646	3085	3445	3713	3885	3964	M
1850	2399	2897	3319	3662	3914	4077	J
1963	2491	2968	3361	3675	3893	4006	J
2399	2855	3240	3525	3705	3793	3759	A
3039	3357	3566	3679	3412	3550	3324	S
3663	3784	3797	3705	3495	3219	2775	O
4127	4056	3880	3608	3240	2784	2252	N
4328	4169	3901	3541	3085	2558	1946	D

of 200 000 lives even though it had been forecast. Periodic droughts such as that which occurred in Sahel between 1969 – 1973 have resulted in the loss of thousands of lives and 80% of the existing livestock. The high variability of monsoon rains also results in frequent disasters; as stated by Smith [33.] "... in some ways life is less hazardous in the desert because the rain there can never be relied upon. The monsoons are trusted and millions have died by this faith".

A substantial part of the population of the so-called 'Third World' lives in tropical regions, by whichever method the tropics are defined. Over 50% of the inhabitants of the earth live in the monsoon areas and the majority of them depend on agriculture for a livelihood. Thus the problem of "... the effects that climate may have upon any effort to increase the productivity of regions characterized by moderately high temperatures and high humidities throughout most of the year ...", as defined by Lee [25], aggravated by the effects of alternating water scarcity and abundance, is one of the crucial problems of tropical water management. As far as tropical agriculture, farming and local industry are concerned, neither traditional nor more modern methods are independent of alternating seasonal surpluses and deficits of water. Besides crop production and animal husbandry, human health and vigour are strongly affected by the tropical climate and the availability of water. Two extreme opinions can be found regarding the nature of these effects; one is epitomized by Huntington [19] as the "... well known contrast between the energetic people of the most progressive parts of the temperate zone, and the inert inabitants of the tropics ..." while the contrasting view is voiced by Cilento [7]; "...first-generation, second-generation and third-generation settlers are performing their work and following their ordinary avocations as they would in temperate climates, and there is at present no indication that the strain of tropical life is an actual one ...".

Falkenmark [8] compared hydrological data for the tropical savanna with corresponding data for the northern taiga zone in order to quantify the differences between ecologically similar regions existing under different climatic conditions. He concluded that a tropical area as opposed to a nontropical area is characterized by lower runoff, higher water deficits, lower water surpluses, smaller groundwater accumulations and considerably greater seasonal variations. Despite the problems of comparing two ecological units exposed to different climate, the conclusions do have some general validity. From another point of view, however, one finds an abundance of different species of birds, mammals and reptiles in the tropical desert compared with less arid neighbouring humid regions. Both animal and plant life forms adapt well to the desert environment and are able to find water or to conserve it in their mode of metabolism.

Deserts and semi-deserts are widely spread within the tropical boundaries. They occur mostly near the extremes of the tropical regions on either side of both the Tropic of Cancer and the Tropic of Capricorn (Fig. 2.2). Studies of the hydrological problems of deserts cannot be confined within the conventional tropical boundaries, but must rather follow the hydro-ecological boundaries regardless of any artificial geographical limits.

Seasonal weather patterns in the central region of the tropics appear to be regular and predictable. On the other hand, when compared with non-tropical regions the tropics show a greater variety of forest species; there is no doubt that with increasing distance from the equator, the monotony of the climate and the variety of plant and animal life decreases, the rainfall pattern in particular becoming more erratic. The amount of rainfall and the time of its arrival becomes more variable, and intense rainfall over a short period may have the same effect as rain of lower intensity falling over a longer period. Because of the high rates of evaporation in the tropics an amount of rainfall of significant proportions for a temperate region may be negligible here.

Mountain ranges also have a significant effect on meteorological and hydrological conditions in the tropics. For example, the Himalayas together with the Sulaiman and Burmese ranges give rise to a climate in India which differs from the climate prevailing in the rest of tropical Asia.

The quality and quantity of the available data for the tropics tend to be poor. Such hydrometeorological and hydrological data that exist derive from regions with many natural and sociological differences; occasionally, political problems are encountered when the data have to be collected from international basins and regions. So far tropical meteorology and climatology, as sources of information on one part of the hydrological cycle, have been in advance of other scientific disciplines concerned with tropical hydrology. With few exceptions, the regimes of

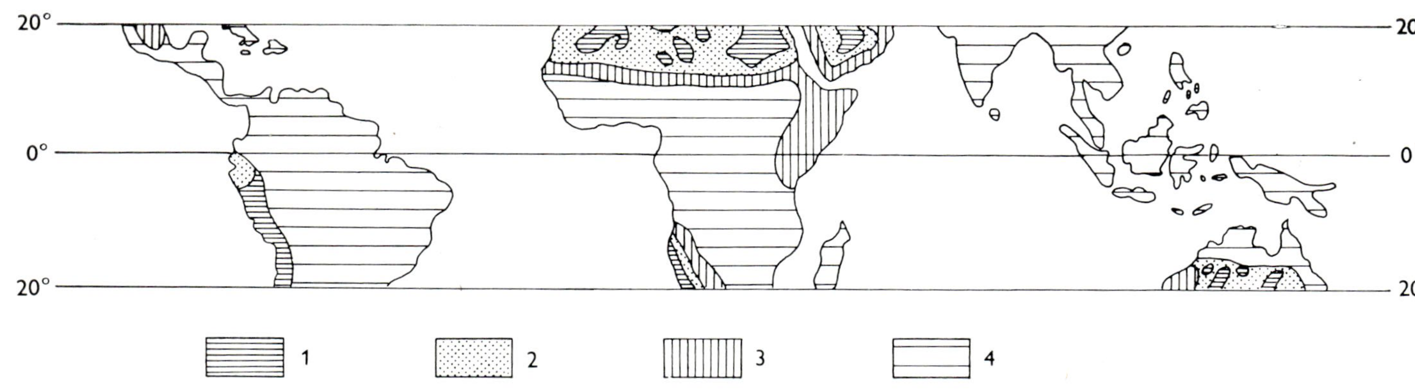

Fig. 2.2 Distribution of deserts and semi-deserts in the tropics 1, waterless desert; 2, semi-desert; 3, acacia semi-desert; 4, other regions

many rivers are still little known. Not much is known about the water regimes of the rocks and soils of the tropics; on the other hand the process of evapotranspiration has become a subject of intensive study in the tropics, because of its importance in tropical agriculture.

2.2 AIR CIRCULATION IN THE TROPICS

According to Tarakhanov [36] tropical air circulation dissipates over other parts of the earth the excess heat which builds up at low tropical latitudes as a result of intense absorption of solar radiation. The difference between the temperature of equatorial regions and that of subtropics is one of the main forces of tropical circulation. Some of the partial circulation such as the trade winds are characterized by a remarkable degree of stability, others such as the monsoons showing a seasonability. Circulation systems on a smaller scale are represented by various types of storms.

Many meteorologists consider the *equatorial trough* to be a main feature of the tropical regions, this trough being a belt of low pressure in the central tropics. Because of the convergence of airflows into the equatorial trough it is referred to as the *Intertropical Convergence Zone* or ITCZ, or occasionally, the intertropical front. Typical of this zone is the cancellation of the opposing effects of the trade winds, also known as the easterlies. Sometimes the equatorial trough is taken to include the entire low pressure belt, while the ITCZ denotes only the active part of it.

Changes in the position of the sun throughout the year are mainly responsible for ITCZ fluctuation. The centre of the ITCZ tends to move up to $10-20°$ north of the equator as it follows the sun into the northern summer hemisphere. The actual distance of the ITCZ from the equator varies with longitude, e.g. in the Atlantic and eastern Pacific the equatorial trough is located near the equator or to the north of it; greatest movement of the ITCZ occurs over Africa and the western parts of the Indian Ocean (Fig. 2.3) where it fluctuates between the Tropic of Cancer (in July) and the Tropic of Capricorn (in January).

Typical of both sides of ITCZ is the reversal of the direction of the wind, and frequently occurring changes in temperature and humidity. Where the ITCZ crosses the coastline, these changes are even more pronounced. The velocity of the wind also varies from place to place along the ITCZ and when the latter is located north of the equator the distribution of rainfall and cloud becomes asymmetrical. In west Africa, for instance, a clear dry type of weather is found north of ITCZ, while south of it for a distance of up to 200 kilometres the cloud cover is moderate and for the next 400 kilometres well developed convection results in heavy rains and storms. Over 600 kilometres to the south of the ITCZ the rainfall decreases to the form of single rainy periods. Fig. 2.4. shows the asymmetrical development of clouds and rainfall along the ITCZ as it appears above west Africa.

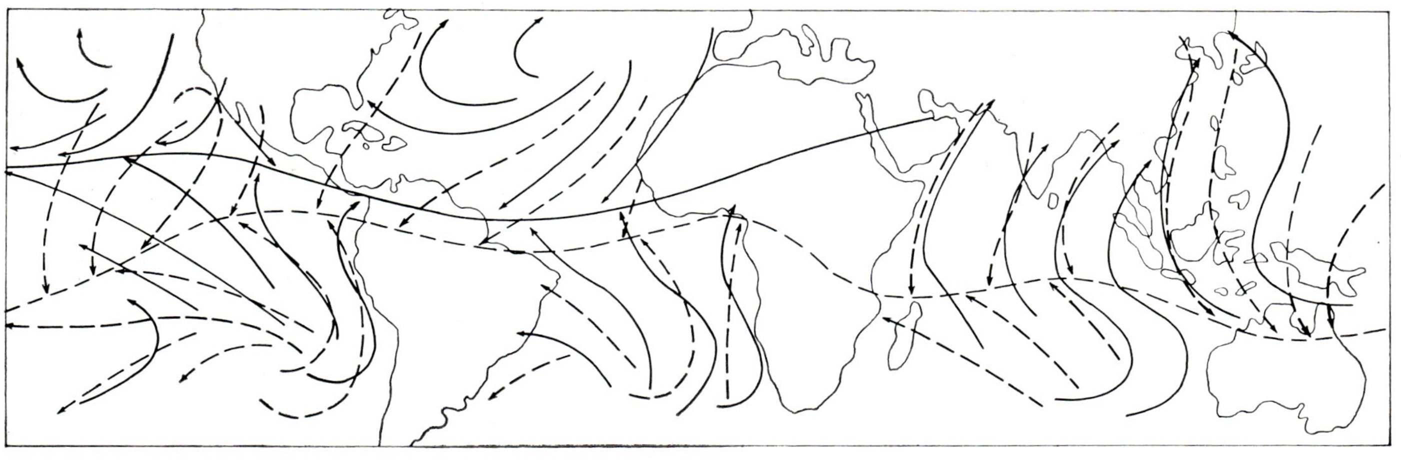

Fig. 2.3 ITCZ in July (full line) and in January (dotted line)

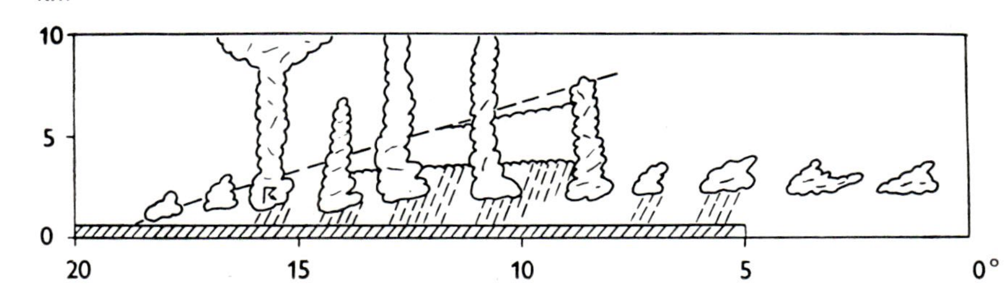

Fig. 2.4 Vertical cross-section of ITCZ in west Africa in August, indicating the asymmetrical development of clouds and rainfall Continental part is hatched

When the ITCZ is near the equator, three zones can be identified within it (Fig. 2.5). In the two outer zones there is a strong convergence with thick clouds and heavy rainfall. In the central zone there is weaker convergence with a prevailing west wind. Between the subtropical belt of high pressure and the equatorial trough, there is the lower troposphere with a prevalent strong airflow, which is southeasterly in the southern hemisphere and northeasterly in the northern hemisphere. This flow is spread over relatively wide belts between a latitude of 20° in the winter hemisphere and 30° in the summer hemisphere, measured from the equatorial trough. More than 30% of the entire earth is under the influence of this airflow all

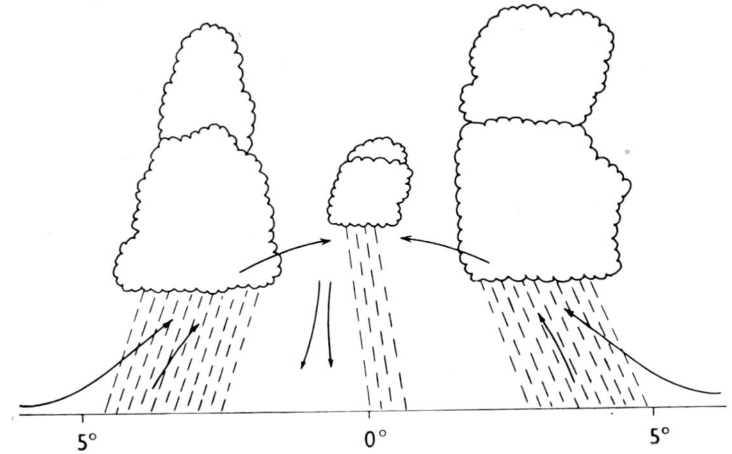

Fig. 2.5 Vertical cross-section of ITCZ when located at the equator

the year round. These so called *trade winds* are stronger above the eastern parts of the oceans than above the western parts of the oceans and the land masses. The trade winds reach a height of about 500 metres in the vicinity of the subtropical high pressure areas and rise towards the equator to about 2500 metres. The air masses of the low trade winds are usually moister and the temperature corresponds to that of the ocean surface waters.

Above these air masses is a layer of inversion (Fig. 2.6.). Its elevation rises in the vicinity of the equator where it reaches a height of 2 kilometres; near the subtropical high pressure cells it reaches a height of only a few hundred metres. The most intense inversion is observed in winter above the eastern parts of the oceans. At higher altitudes is the zone of the upper trade winds which are less strong than the lower ones.

It is assumed that the main force behind the trade winds is the difference in temperature between the hot equatorial belt and the cooler zone of the subtropics. Near the earth's surface, such a difference in temperature results in high pressure towards the subtropics and lower pressure towards the equator. Pressure gradients are responsible for the easterlies diverging in the direction of the equator. Prevailing cool currents over the eastern parts of the oceans cool the lower air layers of the atmosphere and thus the conditions are formed which are favourable for inversion. The increasing instability of the air warmed by the warmer waters of the western parts of the oceans gives rise to vertical air movements. Here the trade wind inversion is lifted to considerable heights and may even disappear.

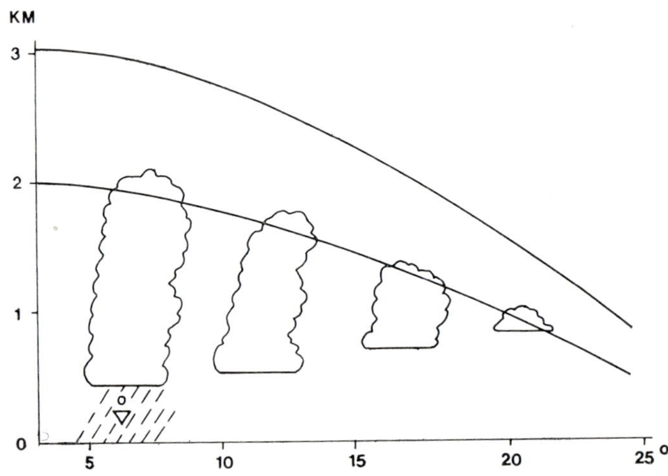

Fig. 2.6 Vertical cross-section of the zone of trade winds. The two curves indicate the location of inversion

The *westerlies* are sometimes considered to constitute the boundaries of the tropics, the name westerly referring to a flow of air originating in the upper troposphere at low latitudes. These winds blow alongside the equator at a height of 8 – 12 kilometres; with increasing distance from the equator their altitude drops, being about 4 – 6 kilometres at a latitude of 25 – 30°. Contrary to the trade winds, westerlies are less regular and show a seasonal variation. They are more pronounced above the western parts of the oceans, particularly in the southern hemisphere and in winter, but their behaviour is much less regular above the continents. A prevailing westerly airflow is observed in the southern hemisphere all the year round between latitudes of 25° and 30° and in the northern hemisphere between latitudes of 30° and 35° only.

The climates of tropical semi-deserts and deserts, especially in the western regions of the continents, are mostly under the influence of *sub-tropical highs* (STHs). The close association of these highs with the subtropical jet stream, and their descent from upper levels, are related to conditions of aridity and semi-aridity on the continents. In the vicinity of the oceans especially, subsiding air may be separated by a layer of cooler air. The low moisture content of the atmosphere above the continents is a result of low moisture advection from the oceans and this allows more radiant energy to heat both the land surface and the air. A low moisture content also gives reduced cloud and rain formation.

While the fluctuation of tropical airflows is more or less regular above the vast areas of the oceans, the situation becomes different in the tropical continental regions. Above the continents the air temperature changes are greater than those that occur above the oceans; the oceans absorb a lot of heat and transfer it to higher latitudes directly whereas heat transfer on the continents can only occur by

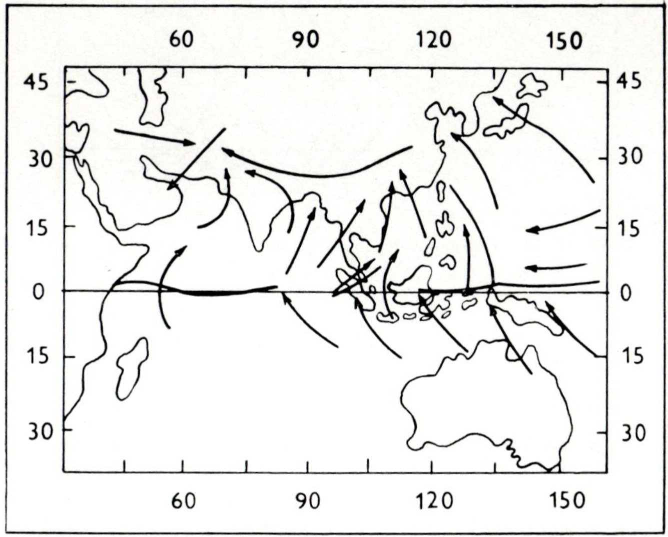

Fig. 2.7 Summer monsoon in Asia, according to Tarakanov

virtue of air movement. In summer, the influx of solar radiation is most intensive between the latitudes of 15° and 25°, and thus the soil temperature in this belt becomes very high. An increase in air temperature contributes to the formation of depressions which are very similar in function to the equatorial trough. The equatorial trough then becomes less pronounced and the trade winds from the winter hemisphere cross the equator from a zone of secondary convergence, the air flowing in the direction of reduced pressure (Fig. 2.7).

In winter the lower temperature of the continents contributes to an increase in pressure with the result that dry air flows from the continents to the zone of convergence which has by then moved south of the equator (Fig. 2.8).

Monsoons are usually assciated only with the Asian continent. The zone of their influence is considerable (Fig. 2.9), because any wind of biannual periodicity and six months' stability of direction can be referred to as a monsoon – in fact the term monsoon refers to the monsoon season. Some monsoons, however, show a pulsation of three to ten days of strong wind alternating with equal periods of weaker airflow.

Deviations from more or less regular circulation are frequently observed in the

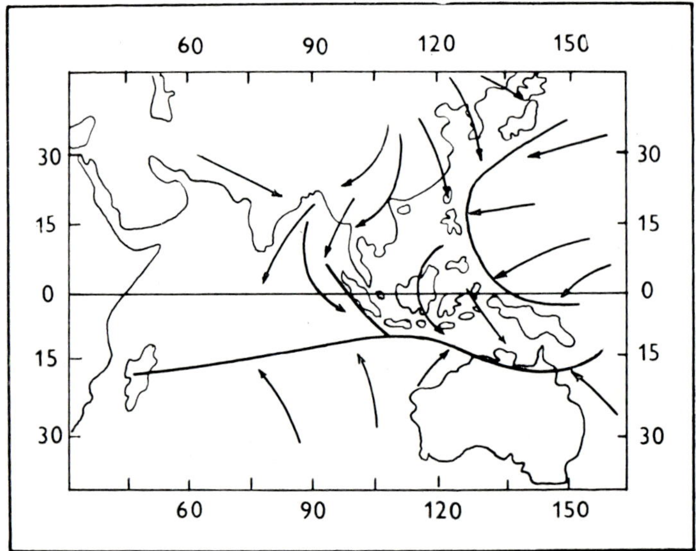

Fig. 2.8 Winter monsoon in Asia, according to Tarakanov

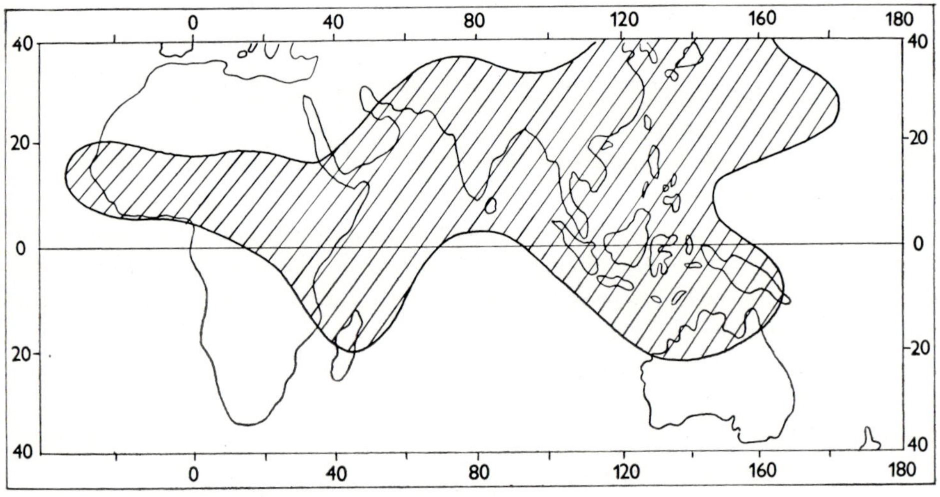

Fig. 2.9 Regions of monsoonal circulation in the tropics

tropics. Such deviations can occur over an area of considerable size, and are generally more intense than similar phenomena occurring at higher latitudes some of them originate in the tropics and then move to higher latitudes. In general they are referred to as *tropical disturbances.*

In principle, we can differentiate between those disturbances that always affect certain regions or show similar behaviour in different regions, and disturbances which originate anywhere in the tropics at random. Both types occur mostly in regions where conditions are favourable for the vertical ascent of air masses, mainly because of the absence, weakness, or great height of trade wind inversion. The areas of most frequent disturbance are the western parts of the oceans and the adjacent land areas, monsoon regions, and the region of the equatorial trough. Generally recognized disturbances include *hurricanes, wave disturbances* and *linear systems.*

Hurricanes, also known in the tropics as *cyclones, typhoons* and *willy-willies* are all easily recognizable on aerial photographs. Initial studies were concerned primarily with the analysis of wind intensity, but more recently, the structure and physical properties of these wind systems have been included in the analysis. Thus one can differentiate between hurricanes having an intensity which is at a maximum close to the earth's surface and decreases with height (so called low hurricanes), and hurricanes with the opposite distribution of intensity (high hurricanes).

The wind speed of low hurricanes varies according to the size of the system. Most arise as a result of synoptic disturbances in the equatorial trough and their origin from high hurricanes is sometimes considered to be a possibility. Their energy of movement derives from the release of latent heat as condensation occurs during the process of convection. The prevailing direction of movement is westward, with a slight deviation towards higher latitudes.

High hurricanes frequently originate at median latitudes and then move toward the tropics, although some originate in the upper troposphere, at low latitudes and remain at an altitude of 6 – 8 kilometres.

Most hurricanes originate in the northern part of the tropical zone of the Pacific where the number of hurricanes arising in the course of the year is about 32, the greatest number developing in September and October. In the northern parts of the tropical Atlantic about 10 hurricanes arise each year, mainly in August and September. In the southwest Pacific there are also about 10 hurricanes a year, mainly in January, but in the Indian Ocean hurricanes seldom occur.

Contrary to the behaviour of hurricanes, *anticyclones* show a clockwise, spiral movement in the northern hemisphere and an anticlockwise movement in the southern hemisphere. In an area of high atmospheric pressure they tend to produce stable weather conditions with frosty winter weather and hot weather in summers. They are, however, less common than cyclones in the tropics. High anticyclones occur more frequently in the southern hemisphere, usually at heights above 5 – 7 kilometres, and are often referred to as dynamic or thermal anticyclones. In dynamic anticyclones no clouds are present, although some clouds may form as a result of weak convection. A typical feature of thermal anticyclones is strong convection and cloud formation at upper and median altitudes.

Low anticyclones are observed mainly in summer in the northern hemisphere in those regions in which the tropical trough has shifted more than 10° from the equator, e.g. the west side of the Pacific areas to the north of the equator, and in the tropical parts of the Atlantic Ocean and West Africa.

Hurricanes are intense disturbances of somewhat small diameter, with minimum pressures of less than 990 mbar, wind circulation speeds of more than 32 m s^{-1}, and torrential rain. Wind systems with circulation speeds of 17 – 32 m s^{-1} are placed in the category of *tropical storms,* while low pressure systems with wind speeds of less than 17 m s^{-1} are referred to as *tropical depressions.* No doubt, the one type of system can easily develop into the other, and many weak systems show a tendency to develop into hurricanes.

Easterly wave disturbances originate on the equatorial side of the subtropical high pressure belt. They occur predominantly at high latitudes and influence the weather in the central troposphere, although they may also affect cloud formation, rainfall and pressure gradients near the surface. Occasionally they may be linked with the initiation of tropical hurricanes. A typical feature of single easterly waves is the presence of two peaks with a single trough. The mean velocity of easterly waves is 20 – 25 kilometres per hour, and the length of the waves is equivalent to about 15° of latitude. Typical examples of such easterly waves are those that occur in the Caribbean.

Equatorial wave disturbances are more typical of the Pacific area. They have an easterly direction and their velocity is below 10 kilometres per hour at the equator, and about 35 kilometres per hour at a latitude of 15°. They can develop as a series of waves of small amplitude, their peaks and troughs occurring along both sides of the equator at the same time.

So-called *linear systems* are synoptic systems, the divergence of which tends to be concentrated along the lower parts of the affected zones, the length of the system being much greater than the width. One of the features of such systems is a strong wind of short duration, as is known for example in western parts of Africa; another is the movement of cool fronts from higher latitudes into the tropics, accompanied by a relatively small amount of rainfall. These systems bring brief periods of bad weather to coastal areas. They occur mainly at the boundaries between the tropics and the subtropics in northern India, southern China, the Caribbean and north Africa.

2.3 RAINFALL

With the exception of monsoonal regions and places where the zones of circulation migrate within latitudes, the tropics are characterized by relatively small transportation of moisture. Seasonal fluctuations of rainfall are relatively stable compared with the situation in non-tropical regions. This holds good except for monsoonal regions and regions where orographic conditions play a dominant role. The rainfall in the regions of high tropical mountains can differ significantly on the windward and leeward sides. An example is provided by the rainfall regime of Kilimanjaro where so-called long rains occur between March and May, and the southeasterly airflow produces maximum on the southern and southeastern slopes while maximum rainfall is brought on the northeast slopes of the mountain by the northeasterly air stream.

Considerably high transport of air moisture and high seasonal variability of rainfall is a typical feature in the zone of equatorial monsoons. In many parts of the monsoonal regions and particularly along the coast and orographic barriers high rainfall has been recorded. Examples are the total annual rainfall of: Cherrapunji, India – 10 913 mm, Mount Waileale, Kauai Island, Hawaii, 11 782 mm, Debundja, Cameroon – 15 000 mm; and Andagoa, Colombia, 7335 mm.

Most of the precipitation comes in summer with the exception of coastal areas where the summer monsoon comes from the continent over the hills as in Vietnam and Somalia, while the winter monsoon comes from the ocean and the rainfall is more uniform.

A considerable amount of rainfall is brought by cyclones to the eastern coastal regions of Malagasy and Australia. The cyclones come more frequently during the spring.

In the trade winds zone and in the zones of high pressure the formation of rainfall becomes more difficult due to the low moisture content of the lower layers of the atmosphere. The trade winds zone has the high pressure of stable anticyclones in the lower layers above the oceans of the southern hemisphere. In summer they protect the formation of rainfall in the trade wind zone because zones of air with a low moisture content are formed above tropical desert areas. This effect of anticyclones is more pronounced in the northern hemisphere in summer and in the southern hemisphere in winter. This phenomenon is connected with the redistribution of air between the two hemispheres. Also, the proportion of land in the southern hemisphere is smaller, and thus the redistribution of rainfall between land and ocean is not as significant as in the northern hemisphere.

Along the western coasts leeward to the trade winds and under the influence of the cool ocean current and the effect of orography desert conditions develop, typical of a great part of the south American and African tropics. This type of climate often reaches deep into the ocean influencing the islands as well. On the other hand in the windward sphere of the trade winds the moist air develops a high amount of rainfall, as in Central America where, however, tropical hurricanes may contribute significantly to the total annual rainfall.

In those parts of the tropics where the situation is favourable for the formation of clouds, high temperatures and moisture produce a considerable quantity of precipitation even from the small clouds.

Other frequent sources of rainfall are weak tropical troughs together with convectional showers resulting from surface heating. The rainfall from these weak tropical troughs is usually more extensive and of longer duration and comes from a uniformly overcast sky.

Thunderstorms, in contrast, are formed locally where the potential energy of latent heat resulting from condensation and fusion in moist air is rapidly transformed into vertical air current kinetic energy. Rapid vertical movement is associated with high surface temperature and convectional reversals. Rainfall of this kind occurs most frequently during the warmer seasons in warmer daylight hours and when the air is sufficiently humid. According to various sources humidity should exceed at least 75%.

During the international project TROPEX – 72 intensive observations indicated that weak to medium rains are most frequent in the tropics, as can be seen in Tab. 2.4. In general, the closer to the equator the higher is the frequency of rainfall from warm clouds, while in temperate regions this type of rain comes from clouds of various temperature.

Tab. 2.4 Occurrence of various types of rainfall, observed in the framework of TROPEX-72

Type of rainfall	Probability No. of events	%
Storms		
weak	23	29.1
medium	20	25.3
heavy	6	7.6
very heavy	9	11.4
Steady rain		
weak	13	16.4
medium	6	7.6
heavy	1	1.3
very heavy	1	1.3
Total	79	100.0

Only recently more information has been obtained on the rainfall pattern in the Pacific region. A strong latitudinal gradient has been found ranging from very dry via very wet to moderate rainfall moving south of the equator. The lowest annual total was recorded at Onotoa (166 mm), the highest at Funafuti (6732 mm). It is remarkable that the two stations are only 7° apart. The lowest record is from 1950 which was a dry year everywhere in the Pacific, the highest record is from 1940. It was observed that the annual maxima for the islands in the dry and moderate rainfall belts are less than the annual average for the islands in the wet zone.

A main requirement for the formation of rainfall is the raising and cooling of large volumes of air. From airborne laboratories 687 clouds were observed in the region of Puerto Rico. None of them reached the altitude of zero isotherm. With the height of clouds at no more than 2300 metres there was no rainfall, at heights up to 3000 metres 20% of rainfall and up to 4000 metres 50% of rainfall occurred.

Beside orographical conditions storms are the main centres for rising and cooling air. However, a large quantity of moisture is required not only because of the water itself, but because of the large amount of latent energy contained in the water vapour.

A typical feature of tropical regions is the rather high spatial and time variability of the annual rainfall. From several atols in the Pacific Ocean annual totals of rainfall significantly less than 500 mm have been reported. It is difficult to establish any regionalisation based on the above records. As far as the annual amount of rainfall and its distribution is concerned, we can find a considerable regularity in the equatorial rainfall with two annual peaks and no pronounced dry season. The converging trade winds from both hemispheres are highly charged with vapour produced by the transpiration of dense vegetation. Apart from the mountainous regions with orographic rain, precipitation is largely frontal and may continue with some breaks for more than a day as steady rain. The moist air above hot ground becomes very unstable and in the later parts of the day convection can become active depending on the height of the tropopause. The rain cools the air and during the night the clouds are often dissolved. Shallow fog dissipates in the morning. Obviously the rainfall pattern is mostly diurnial with some variations.

Rainfall of this type is found in Malaysia, the Philippines, New Guinea, Melanesia, Polynesia, the Congo basin, the northern part of the Gulf of Guinea, eastern Malagasy, extensive parts of the Amazon, parts of eastern Brazil, the West Indies and the eastern coast of Central America (Fig. 2.10.).

Köppen and Geiger [23] considered such a type of rainfall as pertaining to the tropical forest rainfall regions with no less than 60 mm of monthly rainfall. Jackson [21] differentiated between two types, one with at least 2000 mm per annum and 100 mm per month and the other with less than both values but still without pronounced wet and dry seasons.

The monsoonal rainfall regime is regarded as a separate type. The monsoonal rains are among the heaviest on the earth, and they fall within 4 – 5 months in summer. The winter months are almost dry, as can be seen from Fig. 2.11 indicating the rain distribution in Bombay. The annual rainfall is very variable year by year, especially in India and this is very unfavourable from the point of view of agriculture and flood protection.

Monsoonal rain results from the massive condensation produced in the ITCZ by the convergence of the warm and moist monsoon and dry continental air and from condensation above the hot land. Shallow depressions of local origin give long steady downpours and thunderstorms and tropical cyclones increase the

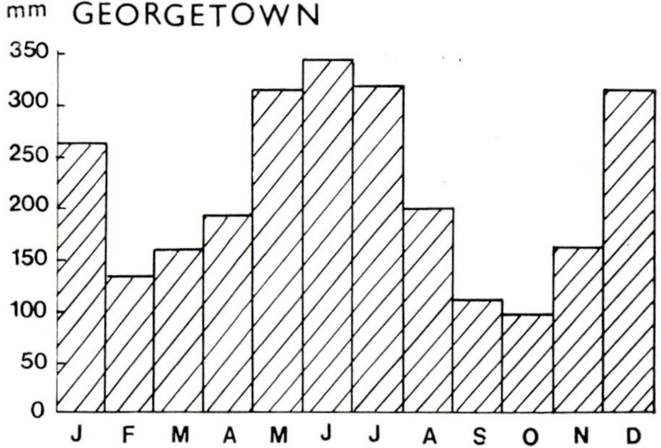

Fig. 2.10 Rainfall distribution at Georgetown, Guyana

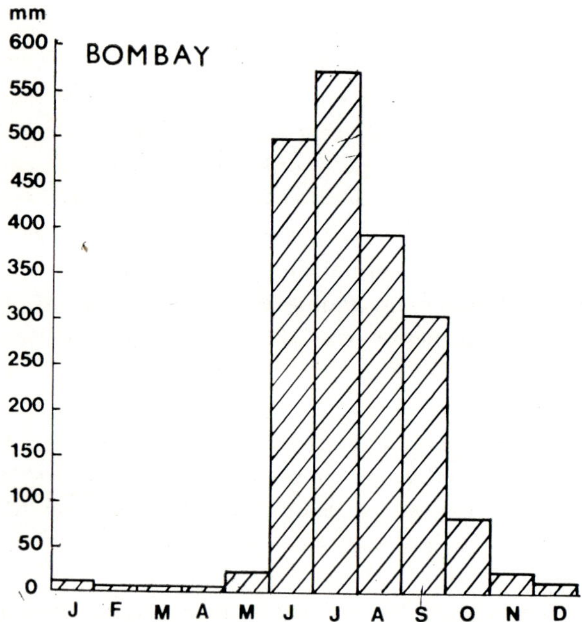

Fig. 2.11 Rainfall distribution at Bombay, India

precipitation in late summer. In India especially orography plays a significant role, because the moist air from the Indian Ocean has to ascend to an elevation of at least 2000 metres. The most striking effect of orography is to be found near the junction of the Khasi Hills and the Arakans where a narrowing funnel causes a rapid ascent of moist air. The rainfall thus becomes enriched from water evaporating from the flood plains in the lowlands. Thus Cherrapanji has a daily maximum of 1040 mm, a monthly maximum of 9300 mm and some sources give the annual maximum in 1860 – 61 as 26 460 mm!

The leeward sides of the mountains in the monsoonal area are shadowed and receive about 20% of the rainfall falling on the windward sections.

Monsoonal rain is typical of a great part of India, Thailand, Vietnam, southwestern Sri Lanka, the western coast of Burma, Malaysia, northern Australia and Sierra Leone.

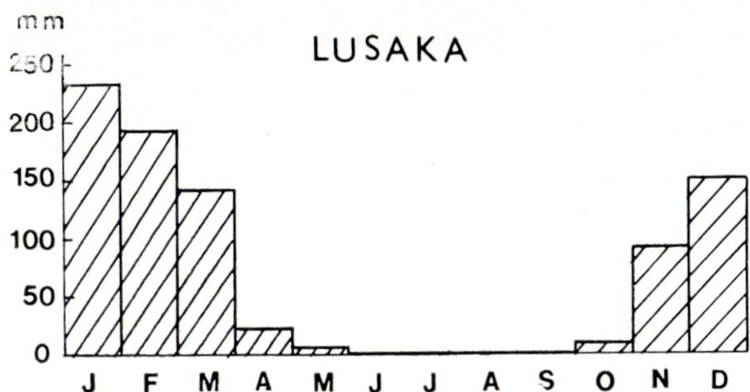

Fig. 2.12 Rainfall distribution at Lusaka, Zambia

A rainfall pattern with periodically dry periods can be found according to Köppen in regions with an annual rainfall of 1000, 1500, 2000 and 2500 mm, and where the driest month has 60, 40, 20 and 0 mm respectively, as a maximum. Perhaps a more typical feature is that the dry periods are sharply limited and the rainy season becomes shorter further away from the equator. The rainy season coincides with the hot season and the hottest weeks are just before the rains come. The summer rain comes after the sun has reached its zenith. The dry period is under the influence of the trade winds. A typical rainfall pattern can be seen in Fig. 2.12. as recorded at Lusaka, Zambia. A rainfall pattern of this type is found in Africa from Ethiopia to Zimbabwe with the exception of the equatorial belt, in western Malagasy, in central parts of Brazil, the northeastern coast of Venezuela, Yucatan and Cuba as well as in India, Thailand and eastern Australia.

The rainfall regime in arid and semi-arid regions is under a greater influence of the trade winds and the main feature is the irregularity of seasonal distribution of rainfall. Köppen considered as semi-arid regions as those which have 700, 600 and 500 mm of annual rainfall at the mean temperatures of 25, 20 and 15 °C respectively. Other authors usually considered slightly lower values as the limits. Sometimes there are differences between two subregions, one with a dry and one with a wet winter period. Such a rainfall pattern is found in the marginal parts of the deserts in southern and northern Africa, large parts of Australia, tropical Arabia, India, coastal Peru and Ecuador. The rainfall pattern at Windhoek, Namibia (Fig. 2.13.) can serve as an example.

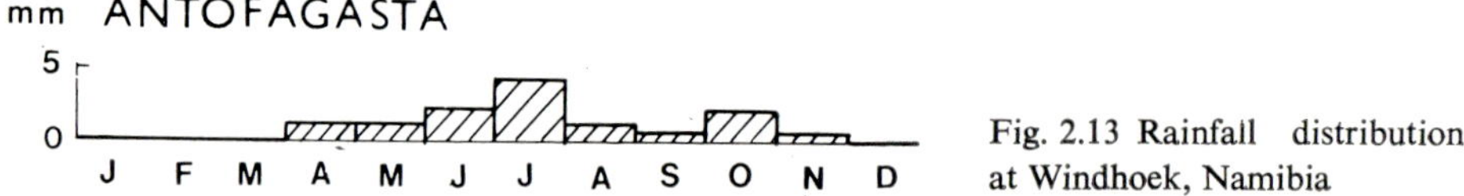

Fig. 2.13 Rainfall distribution at Windhoek, Namibia

The rainfall in the deserts is of less than 350, 300 and 250 mm per year at the mean temperature of 25, 20 and 15 °C, respectively, according to Köppen while slightly modified values can be found in the literature. Extensive regions, such as those in Chile, experience no rainfall at all. Other principal desert areas are in Australia, Arabia and the Sahara and Kalahari in Africa. A record from Antofagasta, Chile (Fig. 2.14.) serves as an example of the rainfall pattern.

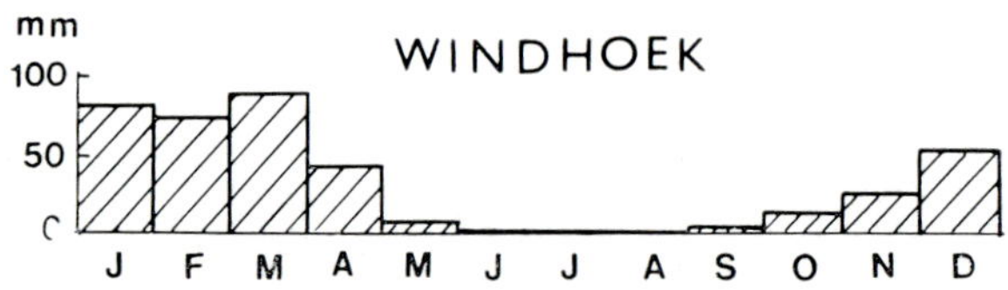

Fig. 2.14 Rainfall distribution at Antofagasta, Chile

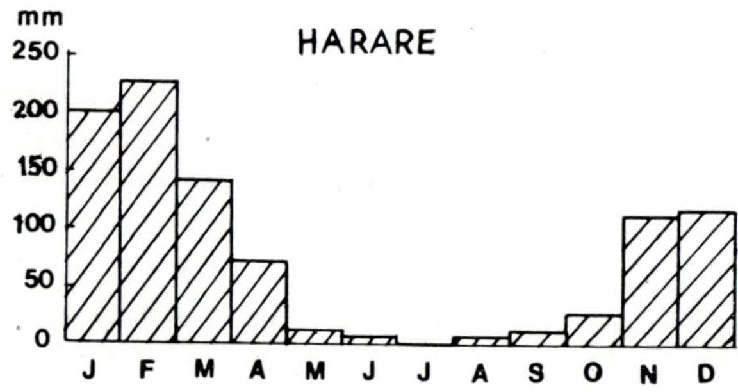

Fig. 2.15 Rainfall distribution at Harare, Zimbabwe

Relatively less extensive regions of the tropics experience warm moderate rainfall. The modification with a dry winter and a wet summer, accompanied by storms is found on the tropical plateau of Central Africa, in southern Brazil, western Mexico and eastern Australia. Köppen considered lowest of annual rainfall to be 300, 400, 500 and 600 mm for the mean annual temperature of 5, 10, 15 and 20 °C, respectively. Other sources consider higher annual rainfall as the possible limit. An example is given for Harare, Zimbabwe in Fig. 2.15.

A rainfall pattern typical of highland or tundra climate can be found on the slopes of tropical mountains in small localities. Actually it is a pattern differentiated by the altitude from the pattern typical of the nearby tropical lowlands. The higlands of Mexico, the area of Lake Titicaca, parts of Bolivia, southeastern Brazil, southern Africa, east Africa and Ethiopia are the regions where the so-called highland rainfall pattern can be found. Also narrow strips of Cordilleras have such a type of rainfall. An example for Addis Ababa, Ethiopia, is given in Fig. 2.16.

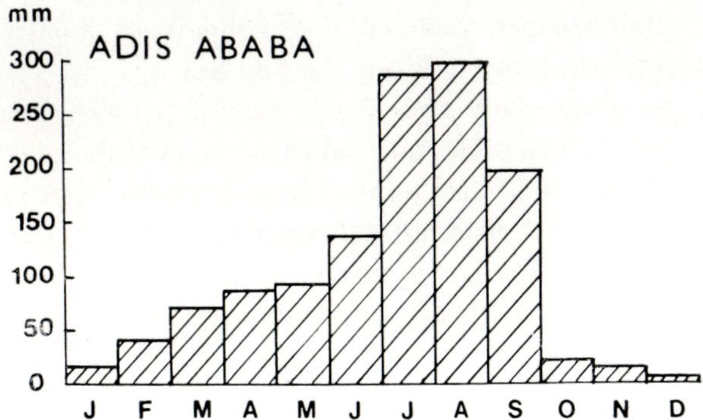

Fig. 2.16 Rainfall distribution at Addis Ababa, Ethiopia

Tab. 2.5 Mean annual precipitation on the continents at various latitudes (mm)

Latitude	Africa	Asia	Oceania and Australia	Americas
30—20 N	37	995		742
20—10 N	550	1350		1395
10—0 N	1388	2650		2535
0—10 S	1320	2595	3050	2010
10—20 S	1070		741	1340
20—30 S	570		323	1020

Fig. 2.17. shows a map of the mean annual precipitation in the tropics. The map is compiled from various sources.

Tab. 2.5. shows the mean annual rainfall for tropical parts of continents, calculated at the Central Soviet Geophysical Laboratory. The numbers combined with the map in Fig. 2.17. provide basic information on the annual distribution of rainfall, however, it has to be noted that small scale regional deviations influenced by various factors can be very different from the smoothed values shown on the map.

For a hydrologist concerned with the various agricultural and engineering aspects of tropical hydrometeorology, the variability of the rainfall pattern at long and short intervals is of equal importance. Tropical rains stem largely from cumulus clouds and heavy rainfall. Diurnial heating leads to afternoon maximum precipitation resulting from convective downpours. For instance, in Bogor, Indonesia, 60% of the total rainfall comes in late afternoon and evening in the dry season of July – September, while in the rainy season it is only 30% of total rainfall. Khartoum in Sudan receives 80% of its rainfall between 18.00 and 6.00 hours from thunderstorms brought by upper winds from the hills bordering the Red Sea.

The distribution of annual, monthly, daily and hourly rainfall is also very uneven even within small areas. In Fig. 2.18 a diagram indicates the distribution of annual rainfall in an experimental area of 16 km^2 on the Central African Plateau.

After his experiments in the agricultural experimental watersheds near Ibadan, Nigeria, Nwa [26] concluded that it would be necessary to have a network density of 1 gauge per 14.77 hectares in order to obtain an accurate picture of mean annual rainfall over the watershed. Similarly Jackson [20] observed in Tanzania

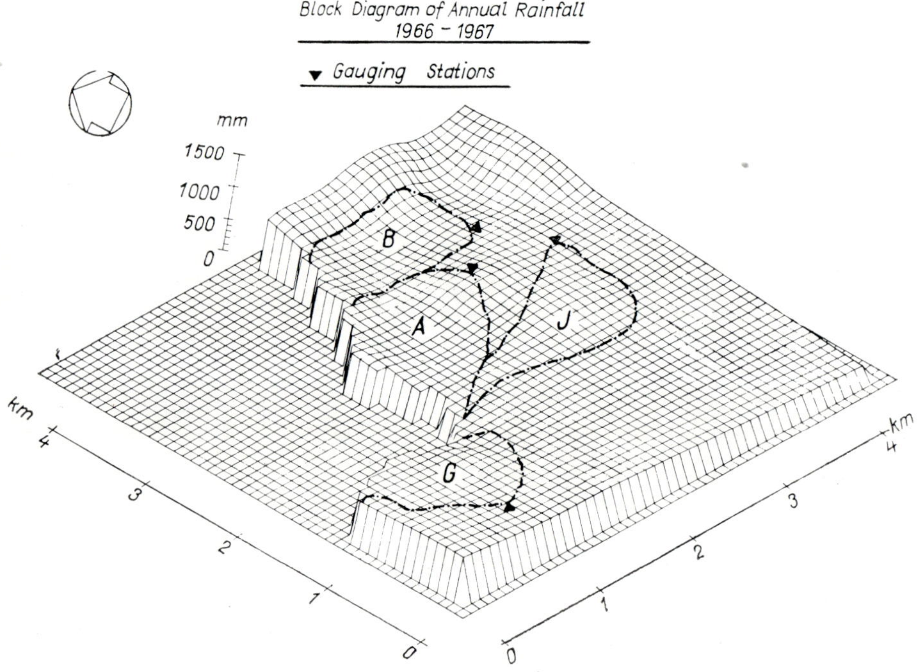

Fig. 2.18 Fluctuation of annual rainfall at the Luano experimental catchments, 12°34′ S, 28°01′ E, 1324 metres a.s.l.

that "even farms only a few kilometres apart may experience very different rainfall in a particular year, although the long-term averages are the same, and permanent influences such as topography are absent".

Lebedev [24] analysed the annual probability of the occurrence of annual rainfall at observation points in tropical Africa. The results indicate high differences in the analysis of the occurrence probability within topographically uniform regions, as can be seen in Tab. 2.6.

Tab. 2.6 Probability of occurrence of annual rainfall (mm) at some stations in the African tropics (Lebedev)

Station	Mean annual rainfall	Probability of occurrence (%)						
		5	10	20	50	80	90	95
Khartoum	158	332	278	222	142	87	69	27
Freetown	3655	4730	4410	4020	3520	3000	3190	2740
Addis Ababa	1266	1750	1695	1455	1215	1045	990	960
Livingstone	722	1040	980	872	720	564	488	430
Windhoek	372	690	610	475	330	255	210	175

At the margins of tropical regions the annual rainfall of low probability differs from the mean annual rainfall to a greater extent than it does near the equator. The data in the table can provide some basic information on the probability of the occurrence of annual rainfall in various parts of the tropics. For some observation points more complex studies have been carried out, testing for the most suitable types of distribution for an analytical approach. It has been concluded that Gauss-Laplace distribution frequently used in temperate regions for such a purpose is not the best for the tropics. Pike [28] proved on the basis of data from Botswana that log-normal distribution was best for the majority of observing points in the Kalahari, however, occasionally some records did not fit into it. An example of the annual rainfall probability analysis for the stations at the margins of the Kalahari Desert is given in Fig. 2.19. Gregory [11] suggested the exclusion of one or two largest totals from the rainfall sequences; this, however, can be done only under the assumption that the values were not correctly measured. The occurrence of low rainfall is more significant in semi-arid areas; the elimination of the highest values can also influence the calculation of low rainfall totals.

Holtz [15] provided a similar analysis for Paraná and for Curitiba located at the margins of South American tropical boundaries. The results of his analysis are given in Tab. 2.7.

Tab. 2.7 Probability of occurrence of annual rainfall at Curitiba (Holtz)

Probability of occurrence (%)	99	90	80	50	20	10	1
Annual rainfall (mm)	830	1080	1180	1380	1590	1700	1950

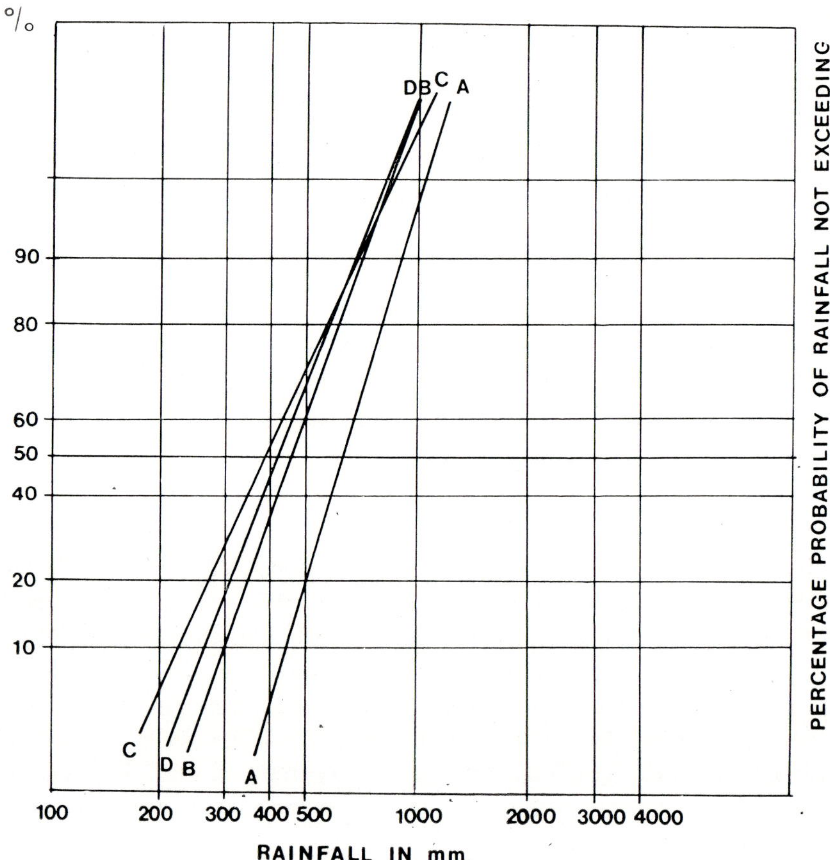

Fig. 2.19 Annual rainfall probability for stations at the margins of Kalahari Desert, according to Pike. A. Kasane, B. Maun, C. Francistown, D. Ghanzi

The daily rainfall regime does not appear to bear a signficant relationship to the regime of the annual rainfall. Daily maximum of rainfall was recorded at Cilaos, Reunion in March 1952. In this case 1870 mm of daily rainfall is far above the daily maxima recorded elsewhere in the African tropics (Tab. 2.8.). This daily record noted only part of an extensive storm during which 2500 mm was recorded in the two days of 15 – 17 March, and 4110 mm in the seven days from March 12 – 19.

Here the effect of the tropical cyclone was combined with a funelling orographic effect in steep hills rising from 0 to 3000 metres above sea level. Normally, however, a daily total of 400 mm is very rare and the previous record was from the year 1911 when a daily rainfall of 1168 mm was recorded at Baguio in the Philippines.

Tab. 2.8 Maximum daily rainfall (mm) in Africa

Month	Country	Daily rainfall (mm)
January	Mozambique	414
February	Mozambique	328
March	Mozambique	386
April	Mozambique	358
May	Ghana	295
June	Ghana	429
July	Sudan	285
August	Senegal	242
September	Nigeria	315
October	Sierra Leone	277
November	Mozambique	259
December	Mozambique	221

In the absence of hourly records which is typical of the greater part of the tropics, any information on daily maximum rainfall can provide useful information in the flood analysis studies. Some hydrologists assume that in most cases the duration of individual tropical storms does not exceed 8 – 10 hours. This may not be true in the tropics. Holtz [16] traced the probability of the occurrence of rains lasting from one to six days at various points in Brazil and found rains that lasted 6 days just as significant as rains of shorter duration. At Paranagua on the Brazilian Atlantic coast he found the relationship

$$h = 90.4T^{0.203}(t + 0.5)^{0.404} \text{ mm}$$

where h is the rainfall total, T is the recurrence interval in years and t is the rainfall duration in days.

For the Iguassu National Reserve in the South American interior the relationship took the form

$$h = 126.0\ T^{0.164}(t - 0.8)^{0.1} \text{ mm}$$

Bailey [3] analysing return periods of daily rainfall on the Central African Plateau found return periods as given in Tab. 2.9. By using empirical methods, Temple and Rapp [37] analysed daily rainfall records in Tanzania. Highest

Tab. 2.9 Return periods of daily rainfall (mm) for some stations of the Central African Plateau (Bailey)

Station	Location	Return periods in years					
		2	5	10	25	50	100
Kasama	10°10′ S 31°12′ E	66	82	91	104	113	122
Livingstone	17°48′ S 25°58′ E	64	89	106	125	141	156
Lundazi	12°30′ S 33°20′ E	56	89	102	119	131	143
Lusaka	15°03′ S 28°30′ E	58	71	79	92	100	109

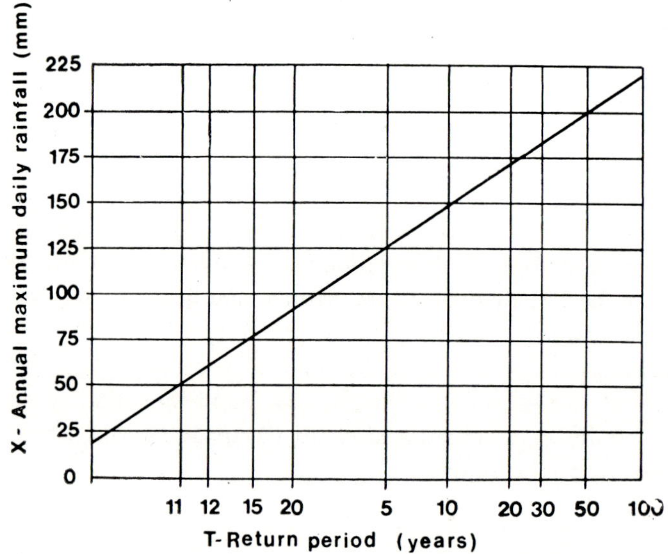

Fig. 2.20 Daily rainfall distribution at Bunduki, Tanzania, according to Temple and Rapp

variabilities were found for Bunduki, as can be seen from Fig. 2.20 Theoretical curves would supply even more extreme values than the empirical ones.

While in his analysis Holtz used Gumbel's distribution, Bailey preferred a Fisher-Tiplett Type I distribution. For annual rainfall analysis Pike applied log-normal and Gauss—Laplace distribution as most convenient. This only indicates a great difference in the rainfall regimes in the tropics and calls for the analysis of rainfall in other tropical regions the regime of which are little known.

The analysis of local storms of short duration is also very important for engineering, design, flood studies, infiltration analysis etc. However, the low density of rainfall recorders in the tropics does not permit wide, analytical work except in small experimental areas. Griffiths [12] developed a formula for the calculation of rainfall intensity I (mm h^{-1}), depending on annual rainfall R, return period d years and duration of the rain t minutes, this being based on the data from Central African Plateau and for annual rainfall up to 1000 mm:

$$I = \frac{26(0.16R - 83)(\log d + 2) + 10}{t + 15} \quad \text{mm h}^{-1}$$

However, it is not entirely clear whether any relationship exists between annual rainfall and storm intensity.

Jackson [21] provides an example from Mozambique where at a station with an annual total of 300 mm, 100 mm was recorded in one day, Oliver reported a daily total of 111.5 mm in Port Sudan, where the long-term annual total is only 110 mm.

Balek [5] developed a formula for the region on the Central African Plateau

$$I = e^{(-1.097 \log t + 2.07)}$$

where I is the rainfall intensity (mm h^{-1}) and t is the duration of rainfall (min). The formula is valid for storms with 25% probability of occurrence. For a 1% storm probability another formula was developed:

$$I = e^{(-0.94 \log t + 2.55)}$$

A comparison of both curves can be found in Fig. 2.21.

Souza found a formula for Southern Brazil [35]:

$$I = \frac{99.154F^{0.217}}{(t + 26)^{1.25}}$$

where I is the rainfall intensity, mm min^{-1}, t is the rainfall duration in minutes and F is the return period in years.

In the region of great African lakes Raman [31] found the following relationship between rainfall intensity and rainfall duration:

$$I = 26.84(t + 0.4)^{-1}$$

where I is the rainfall intensity (mm h^{-1}) and t is rainfall duration in hours.

One of the most comprehensive works on rainfall intensity analysis in the tropics was accomplished by Babu et al. [1]. They divided India into five regions and analysed the records of each separately. For Indian tropical margins they found the relationship

$$I = \frac{7.4645T^{0.1712}}{(t + 0.75)^{0.9599}} \quad \text{cm h}^{-1}$$

and for Indian southern tropics

$$I = \frac{6.311T^{0.1523}}{(t + 0.5)^{0.9465}} \qquad \text{cm h}^{-1}$$

T is the return period in years and t duration in hours.

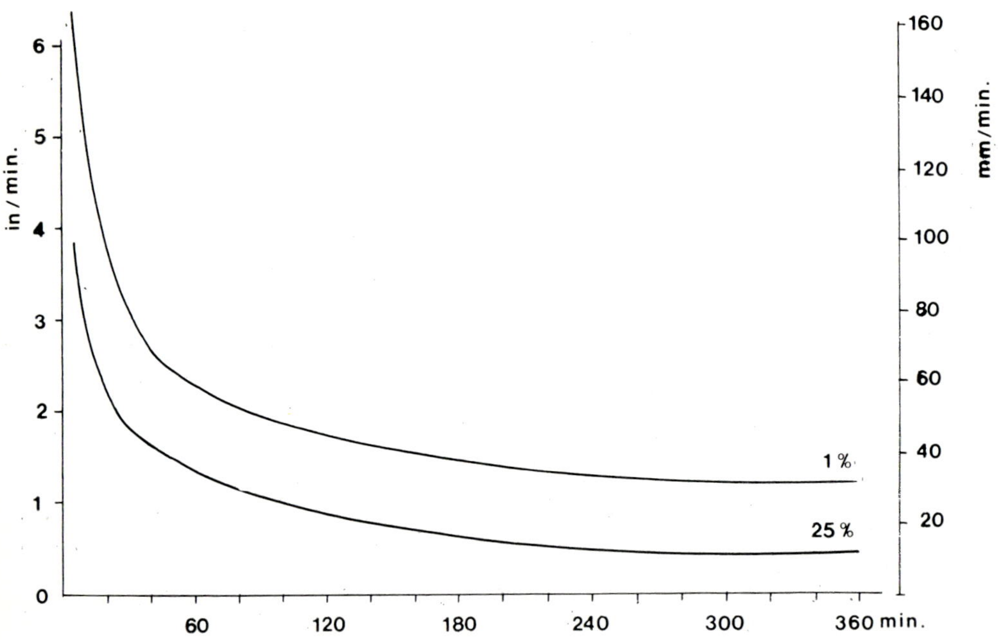

Fig. 2.21 Rainfall intensity-duration curve for storm probability of 1% and 25 %, at Zambian—Zairean border

Perhaps it would be possible for more formulae to be compiled. However, as can be seen from earlier ones the results only indicate how difficult it would be to introduce some regional factors in order to develop a formula generally valid for tropical regions. It is likely, that the orographic pattern with the topographic conditions of the whole region and some other factors could play a significant role. Therefore a laborious method based on the analysis of records available for the specific area being studied, can only be recommended for any tropical region.

Henry found that tropical rains of more than 10 mm fell in south-eastern Asia and parts of South and Central America in intervals of 1 to 3 hours and in no more than one storm a day. Hjelmfelt [14] studied the rainfall pattern in wet and dry seasons in the Amazon basin. From the analysis of records on the Atlantic coast and in the western part of the basin he found no significant difference at various observation points. The results indicate that rainfall is not continuous throughout the rainy season, and does not necessarily fall every day. Fig. 2.22 indicates that 50% of the intervals between rainstorms are of 6 to 10 hours duration or more and therefore it rains from two to four hours each day. Even during the wet season there are four days in a month without any rain.

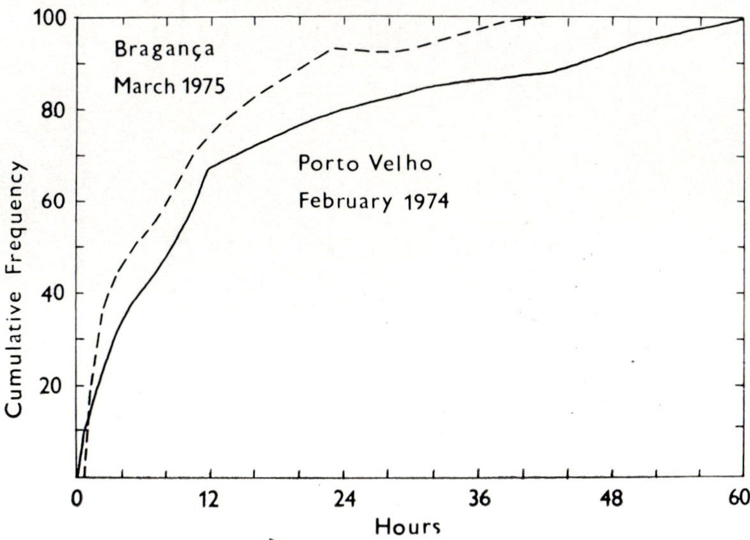

Fig. 2.22 Interval between storms during the wettest month at Bragança and Porto Velho both in the Amazon basin, as a percentage. According to Hjelmfelt

A very comprehensive study on rainfall duration and frequency was carried out for Sri Lanka by Baghirathan and Shaw [2]. They studied the convectional and monsoonal types of rainfall pattern with an annual total of 967 to 5461 mm. Derived relationships are of the regional type dividing the country into six regions. A comparison of the relationships developed for the lowest and highest values is given in Fig. 2.23. It indicates a great variability in the depth – duration – frequency relationship of the rainfall. The results clearly reflect the impact of the monsoons and topography of the island.

The extent of the rainstorms is vitally important to agriculture, culvert design and erosional processes. The lower limit, which could be considered in temperate regions as a flood producer may be of little importance in the tropics, being consumed by evaporation. On the other hand, the maximum limit can result in catastrophic soil erosion. Riehl [32] considered the lower limit defined above as any rainfall of 5 mm while a threshold critical value for Africa was given 25 mm h^{-1} (Hudson, [18]).

According to Griffiths, in the central Sahara a rain 174 mm h^{-1} was recorded for a duration of three minutes, and at Golea 92.4 mm h^{-1} for 25 minutes. At Beni Abbas 46 mm h^{-1} rain was recorded for a period of 63 minutes.

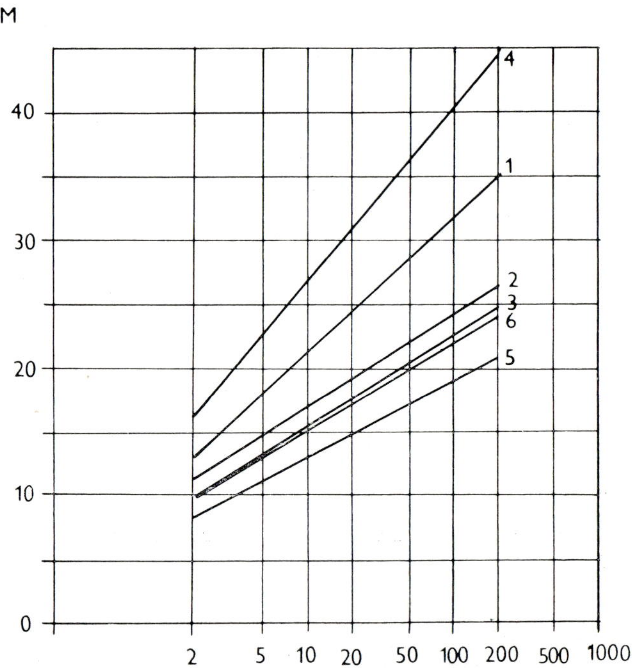

Fig. 2.23 Rainfall depth-frequency relationship for 24 hours duration, Sri Lanka; 1, east; 2. northwest; 3, southwest; 4, south; 5, central; 6, south-east. According to Baghirathan and Shaw

Also very little is known about the spatial variability of rainfall. For an objective analysis to obtain depth – area – duration curves, there is a need for a great number of rainfall recorders in the region studied. Therefore very few results are available from different parts of the tropics.

Tab. 2.10 Distribution of the rainfall (mm) in Nvando basin, Kenya (Raman)

Date of storm	Area (km^2)				
	10	100	1000	2000	5000
27/11/1961	83	80	65	53	34
27/4/1965	58	46	35	30	22

Raman, in a survey of the World Meteorological Project found in Nyando basin, Kenya, the distribution of rainfall within an area of 5000 km^2 as given in Tab. 2.10. The analysis was based on single rainstorms only [31].

Holtz [17] found a linear relationship between the area and total rainfall of varying duration as indicated in Fig. 2.24.

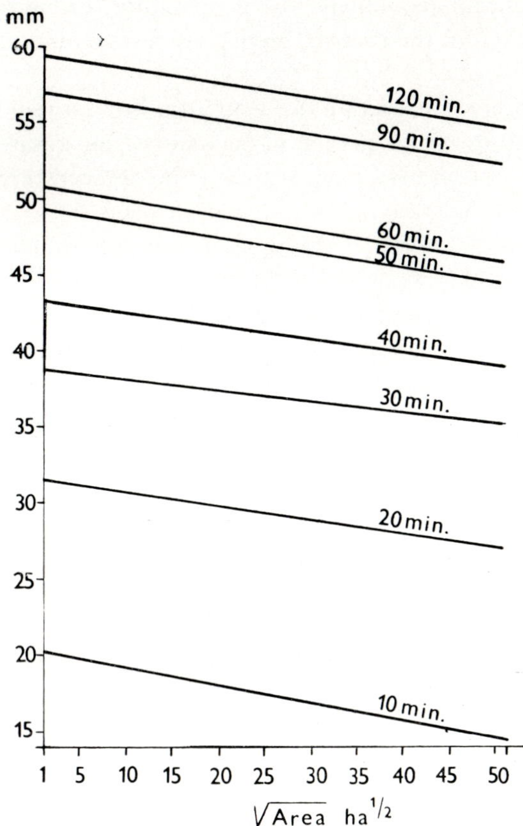

Fig. 2.24 Relationship between the rainfall total and area of storms of various duration, Paraná, Brazil, according to Holtz

Balek [5] developed an equation for the depth – area – duration relationship for the Central African Plateau:

$$P = te^{(-1.097 \log t + 2.07)} - 4.035A \log t + 0.743^{\;(t+3.7)} A^2$$

where P is the mean annual rainfall depth (mm) t is the rainfall duration (min) and A is the area (km^2). The curve is valid for the recurrence interval of four years. Shortage of records prohibits further extension of the results.

2.4 TEMPERATURE AND HUMIDITY

A dominant factor in the temperature regime of the tropics is the radiation balance. However, some other factors also play a significant role. A positive balance of solar radiation resulting from an excess of incoming radiation all the year is responsible for high air temperature. Because solar radiation is influenced by astronomical factors with a significant periodicity, the temperature regime is also highly periodical and random parts in the thermal regime are less significant than outside the boundaries of the tropics.

Considering the variability of monthly temperature in accordance with latitude, one can see that the isotherms indicating the fluctuation of maximum temperature are more or less identical with ITCZ. In January the isotherm of the temperature maximum is at a latitude of 5° south of the equator and almost identical with ITCZ, while in July it is found about 20° north of the equator and thus slightly deviated in the northern direction from ITCZ.

The northern tropics are warmer than the southern tropics by 2 °C and also the amplitude of the temperature change within a year is greater in the northern than in the southern tropics. At a latitude of 30° north the amplitude is 13 °C, while in the latitude of 30° south it is only 7 °C. Also the maximum of mean annual temperature is most frequently found along the latitude of 5° north, where it becomes

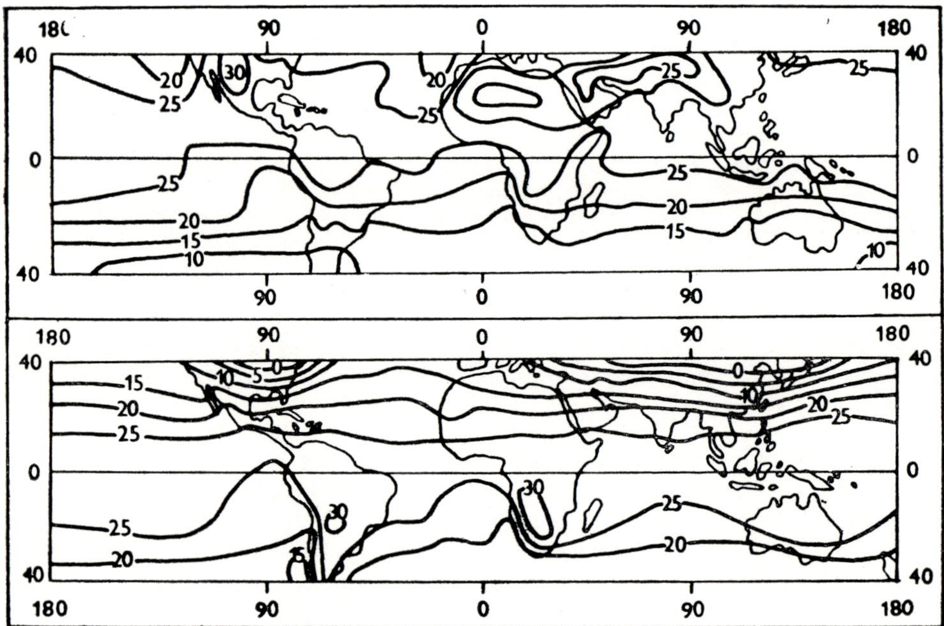

Fig. 2.25 Air temperature (°C) at the sea level in July (upper graph) and in January (lower graph)

almost negligible. The isotherm of maximum monthly temperatures is, however, deviated to higher latitudes above continents in the summer hemisphere and to the equator above the ocean. This can be seen in Fig. 2.25.

Another interesting graph was developed by Riehl [32] for the centre of the southern tropics. It indicates the variability of mean temperature in summer and winter (Fig. 2.26.). The decline of the mean temperature along the west coast of Africa and South America is explained by the transfer of cool air from the higher latitudes and the ascent of deep, cool water along the coast. Another marked feature is a lower fluctuation of temperature over the oceans than over the continents. The highest temperatures are found in summer over Africa and Australia.

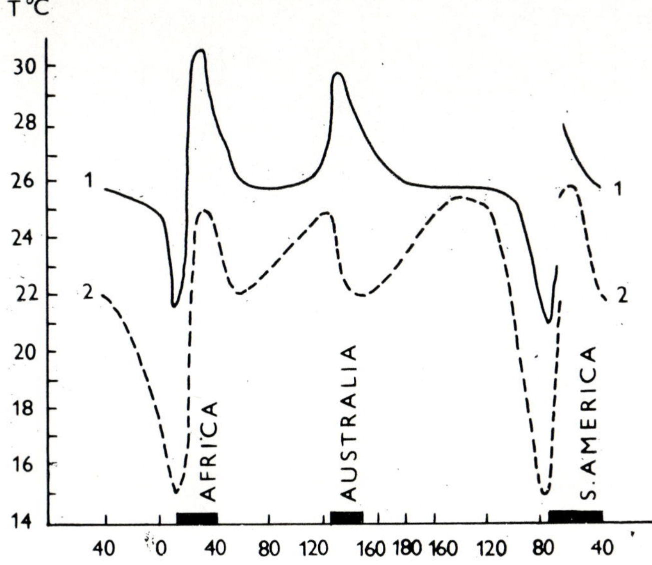

Fig. 2.26 Temperature variability at sea level according to Riehl; 1, in summer; 2, in winter, both 15 °S

Annual variability of temperature as found in the tropics, is of the oceanic or of the continental type. The oceanic type is found over the oceans and along the coastal regions over the land. Maximum temperature is recorded in late summer and minimum temperature in the middle of winter. The temperature decreases with increasing latitude (Fig. 2.27.). The temperature fluctuation at Havana, Cuba, 23° N, where the monthly maximum is in August and minimum in January serves as an example. At Dominica in the West Indies, 15° 25′ N, the monthly temperature remains more or less stable at 32 °C from May to October. At Georgetown, Guyana, 6°40′ N it remains stable throughout the year at about 30 °C. From the theoretical point of view, two maxima and minima should be found at the equator corresponding to the position of the sun, however, cloudiness and rainfall cancel this effect.

Much less is known of the vertical variability of temperature in the tropics. The results of the temperature measurements conducted in the framework of TROPEX 72 and 74 in the ITCZ zone, and in the zone of the trade winds can be seen in Fig. 2.28. The highest temperature gradients of 8° per 100 metres were observed above the inversion in the trade winds zone.

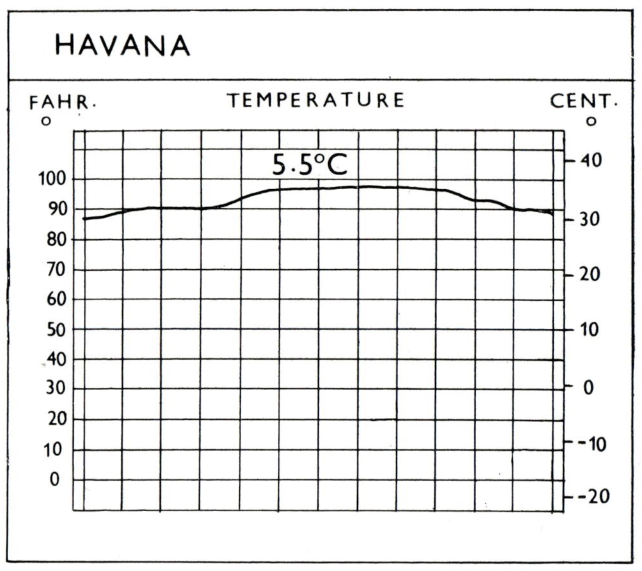

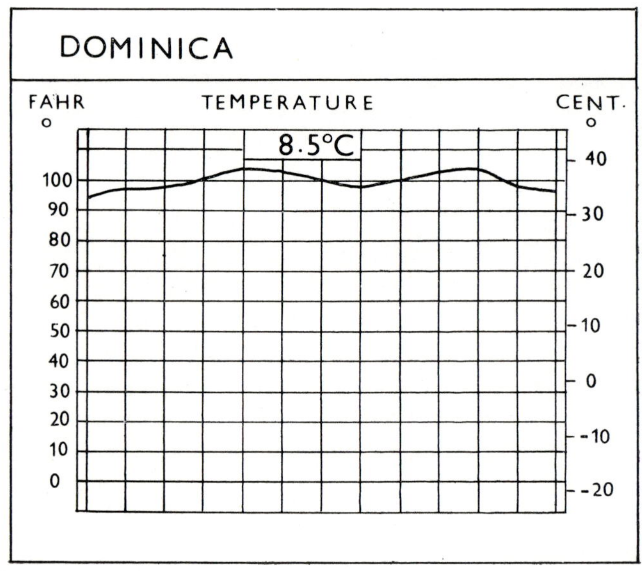

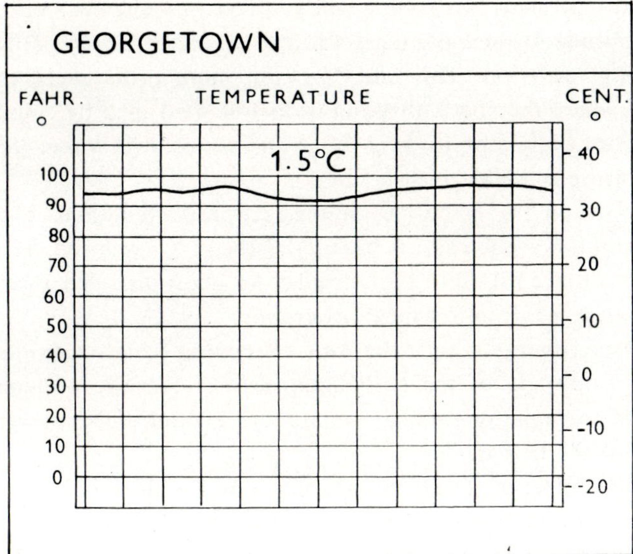

Fig. 2.27 Comparison of the monthly means of daily maximum temperatures and the mean annual range of temperature at Havana, Cuba (23° N), Dominica, West Indies (15° 25′ N) and Georgetown, Guyana, (6° 40′ N)

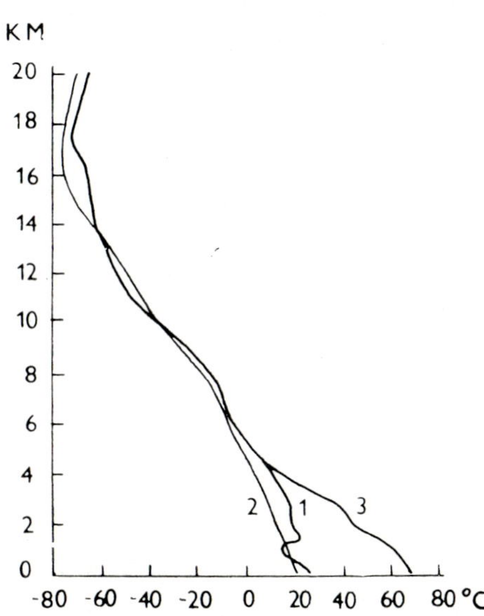

Fig. 2.28 Vertical profile of air temperature in the tropics, according to TROPEX 72. 1, Trade winds zone; 2, ITCZ zone; 3, Khartoum

From the hydrological point of view daily temperature variability is equally important. Here the main role is played by the location of the observation point. Above the oceans the daily temperature amplitude is very small. With increasing distance from the ocean to the inland area the temperature increases, however, the presence of inland lakes, swamps or mountains can have a strong effect on the regime. Again, there is a difference between the distinctive continental and oceanic patterns. In the continental pattern an additional role is played by altitude. With increasing altitude the amplitude of the daily temperature also increases according to the increase of incoming radiation. This effect is even more pronounced in regions with a dry climate where the effect of radiation is not weakened by cloud cover. An indirect decrease of daily amplitude is found in deep valleys where the cool air lowers the temperature at night.

A different situation exists on the slopes of isolated tropical mountains. The slopes are open to the circulation of air and the heating of the soil is reduced. The temperature gradient decreases according to the altitude. An additional effect can be produced by clouds formed as a result of local orography.

Dense clouds above the continents reduce the rate of incoming and outgoing radiation and reduce the amplitude of the daily temperature. Because of high cloudiness in the vicinity of the equator the daily temperature amplitude is lower than on the poleward side of the tropics.

An example of the seasonal effect of cloudiness was given by Kendrew [22] who compared the fluctuation of air temperature in Manila in the dry and rainy season. The amplitude increased during the rainless period.

Air humidity is highly dependent on air temperature and air pressure fluctuation. In the tropics two types of air masses are formed, one the continental tropical air (CTA) and the other oceanic tropical air (OTA). The poleward sides of the tropics can be under the influence of air masses from higher latitudes which can influence the so-called normal humidity regime of the tropics.

Centres of the formation of continental tropical air above the dry regions are characterized by low humidity. In summer, the temperature rises over 40 °C, in winter over 30 °C, with a vertical gradient 0.6 – 0.65 °C per 100 metres. The humidity of continental tropical air in summer is at earth level at 16 – 18 % and increases to 24 – 26 % at an altitude of 3000 metres. Above that level it decreases again. In winter the humidity is at earth level at 24 – 26 % and at an altitude of 1000 to 2000 metres it is lower, while above this it rises and at an altitude of 4000 metres it reaches 35%.

Near the coastal regions humidity rises. In the vertical direction the maximum humidity rarely goes beyond an altitude of 3000 metres. The humidity of oceanic tropical air is highly dependent on regional factors. In the ITCZ high humidity reaches higher altitudes. Air humidity in the zone of the trade winds is more variable. In lower layers the maximum is found at an altitude of 1500 – 2000 metres, while in the upper layers there is a second maximum at an altitude of 6000 – 7000 metres. Between these two zones there is rather dry air. Oceanic air after it has reached the continent is transformed in such a way that the humidity decreases. It still reaches 50 – 60 % however.

Daily variability of air humidity can have a very high amplitude, especially when disturbances occur. Fig. 2.29. records the air humidity, temperature and air pressure in one day, resulting from the occurrence of a cyclone at Manila. The record was compiled by Kendrew [22].

At this point one should recall the effect of water which condenses on clear calm nights as dew. This type of atmospheric water can constitute a significant additional source of water for plants and for man. Pereira [27] observed herdsmen in Tanzania collecting water at dawn by sweeping the dew from tall grasses. The effect of dew should be studied more intensively in the tropics because it appears that under favourable conditions the amount of dew can equal potential evaporation in the same interval.

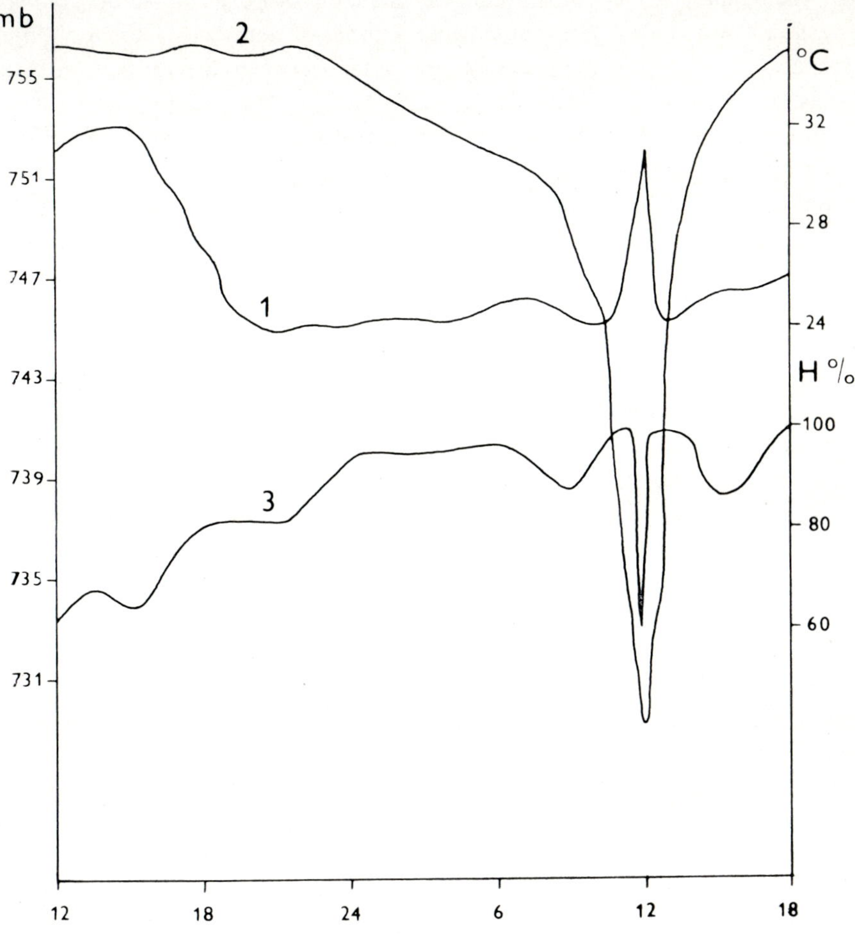

Fig. 2.29 Daily fluctuation of air temperature (1), air pressure (2), and humidity (3), at Manila, resulting from the occurrence of a cyclone at Manila. According to Kendrew

2.5 CLIMATE

Special features of climatic conditions exist in all tropical part of the continents and to some extent of the oceans as well.

Africa has the largest continental mass in the tropical belt. The continent is symmetrical to the equator, however, the rainfall distribution is somewhat distorted. Along the equator in the region of the equatorial trough there are a number of rainy days and the temperature regime has a low amplitude. The belts of trade winds in the northern hemisphere and in the southern hemisphere reach beyond the tropics and so at the poleward sides of the tropics a rather low amount of annual rainfall is typical. The dominant feature in the rainfall asymmetry is seen above the Sahara. A marked difference in the climatic pattern also occurs between the east and west coasts. The temperature regimes of both coasts are very different, which accounts for the temperature difference between the Indian and Atlantic Oceans.

Annual rainfall along the equator is frequently about 2000 mm and is marked by two distinct peaks, typical for equatorial belts in keeping with the equinox of the sun. In this belt the rainfall decreases from the western to the eastern coast.

The temperature amplitude, small at the equator, increases both southward and northward. Here the differing impact of the oceanic and continental regime is very pronounced. Only one peak in the annual rainfall pattern exists in the region of the trade winds. With prolonged dry season on the poleward sides the regime goes over to a semi-desert and desert type of climate, such as exists in the Sahara and Kalahari. High subtropical pressure with clear and dry weather prevails in these regions, particularly in winter.

The surface of the deserts is heated to 80 °C and this influences the temperature of the air which reaches as much as 50 °C. Radiation from the earth's surface during the night cools the lowest air layers to such an extent that at a latitude of about 30° the temperature can fall to below 0 °C and in the highlands on the Central African Plateau this can be experienced at even lower latitudes.

Owing to different temperature of the ocean currents and several types of prevailing winds, the southern parts of the continent have a variety of climates. The Atlantic coast is cooler with little rainfall. Deeper inland rainfall peaks come in the summer period.

The snowline on high African mountains lies between 3875 and 4750 metres.

In Central America the climatic regime is greatly influenced by the orography. Four different types of climate can be distinguished. From the coast to an altitude of about 1000 metres there is a maritime climate with more or less constant temperature about 23 °C. Up to a height of 2000 metres there is a mean temperature of about 17 °C with rather low amplitude of seasonal variability. Still inhabited but rather cooler belts exist along an altitude of 3000 metres with the mean temperature below 17 °C. Above is a cool belt with occasional frost and rainfall prevailing during spring, autumn and winter.

The Atlantic coast is influenced in winter by northern air masses, and in summer by air from the east. Late summer has rather high rainfall, in Yucatan rain prevails in the autumn.

The Pacific coast is drier than the west coast and has prevailing east winds in winter and west winds in summer.

A great part of South America belongs to the tropics. A typical feature of the continent which has an impact on the climate is the mountain belt of the Andes which divides the narrow coastal strip from the plateaux and slopes with variable climate. Here the snowline is at an altitude of 4200 metres.

Tropical regions have a distinct tropical climate with low seasonal and daily temperature amplitudes and a daily maximum early in the afternoon. The coastal regions of Colombia, Venezuela, Guyana and Brazil are under the influence of trade winds all the year round and the annual rainfall reaches 2500 mm. The weather is warm, humid and rainy.

The hinterland as in the Amazon basin receives moist air from east winds with prevailing low pressure from December to April. A double peak of rainfall seasonability is experienced only in higher latitudes. In summer the rains shift southward. The climate of the plateaux of the Andes is often described as eternal spring. Night and mornings are rather cool. Cloud cover increases in the afternoon and frequently in storms.

The Pacific coast of Colombia north of the equator corners into the region with the rainfall maxima. In contrast, the coasts of Ecuador, Peru and Chile are influenced by a cool Peruvian current, which brings low clouds, low air temperatures and slight rainfall. Thus this part of the region is dry. The mountain chains prohibit the movement of maritime air into the hinterland. Eastern slopes and coastal plateaux experience little or no rainfall. The poleward sides of the American tropics have rather low rainfall being under the influence of high subtropical pressure.

In the climate of tropical Asia the monsoonal regime is dominant. During winter high pressure, the dry winter monsoon descends towards the Indian ocean and the climate from October to the end of February is clear, dry and mild. The climate of southern China is cooled in winter by the winter monsoon. The climate here is cooler than at the same latitudes elsewhere. The air blowing from the continent becomes moist over the ocean and brings rain to the Phillipines and to the islands of tropical Asia.

In spring and summer the lowering of air pressure above Asia produces a reverse movement and the moist air brings monsoonal rainfall to the Asian continent, between June and October. A secondary peak of rainfall lasts until December and typhoons occur in the oceanic part of the region.

In the equatorial islands a dominant role is played by the orography and rainfall of orographic origin occurs all the year round. This rainfall reaches its maximum in winter in the southern parts of Sumatra, Borneo and Java while in the northern parts maximum rainfall comes in summer. These islands are also under the influence of the northeastern monsoon from the Australian continent.

Australia is the driest continent. With the southward shift of STH in summer the monsoonal regime begins to influence the continent. In northern parts of Australia one can differentiate between the summer rainfall of October to April and the dry period in winter between May and September. Mean annual rainfall varies around 1000 mm and rapidly decreases in the interior, so that the central part of the continent is of semi-desert and desert type. The west and southwest coast of the continent receives winter rainfall while the east coast is moistened in summer by rainfall from the Pacific. The amount of rainfall is highly variable.

Owing to the influence of the ocean the northern part of Australia has a mean temperature of 24 – 25 °C. while in the interior it can reach 55 °C. Here, the thermal amplitude is very similar to that of the Sahara.

For the Pacific islands maritime oceanic climate is most typical. Under the additional influence of the trade winds the climate is very pleasant. The typical annual amplitude of the temperature is 0.4 – 0.5 °C and never exceeds 5 °C. Most of the rains come in summer, but some rains come all the year round and there is no pronounced dry season except in flat islands where the orography does not contribute to the ascent of hot moist air. On mountainous islands the windward sides are richer in precipitation than the leeward sides.

It is obviously very difficult to classify tropical climate in great detail. Nevertheless, many attempts have been made to classify them. Voeykov [39], Forsberg et al. [9], Pollock [30], Köppen [23], Herbertson [13], Trewartha [38], and others have attempted to produce various types of classification. From the hydrological viewpoint, Köppen's classification probably gives the most comprehensive account of the variety of climatological regimes in the tropics. The following are recognized:

Tropical rain climate (*A*) which has a mean annual temperature of above 18 °C all the year round. With the temperature at 20 °C at least 600 mm and at 25 °C at least 700 mm of the annual rainfall should be available. Three subtypes are defined:

- *tropical rainfall forest climate* (*Af*) in permanently wet regions with some rain in all months,
- *monsoon climate* (*Am*),
- *periodically dry savanna climate* (*Aw*). Here for an annual rain 1000, 1500, 2000 and 2500 mm the driest month should have 60, 40, 20 and 0 mm of precipitation or more, respectively.

Dry climate (*B*) is found in the tropics in two subtypes:

- *steppe climate* (*Bs*), where at the mean annual temperature of 25, 20 and 15 °C maximum annual rainfall is 700, 600 and 500 mm, respectively.

- *desert climate* (*Bw*), where at the mean annual temperature of 25, 20 and 15 °C maximum annual rainfall is 350, 300 and 250 mm, respectively.

Warm rainfall climate (*C*) in which the temperature of the coolest month is between 3 and 18 °C and at the mean annual temperature of 5, 10, 15 and 20 °C the mean annual rainfall is more then 300, 400, 500 and 600 mm, respectively.

The following subtypes are found in the tropics:

- *warm climate with dry summer* (*Cs*),
- *warm climate with dry winter* (*Cw*),
- *warm wet climate with precipitation in all months, temperature in the warmest month below 18 °C and at least four months over 10 °C* (*Cfb*).
- *warm wet climate with precipitation in all months, temperature in the warmest month over 22 °C* (*Cfa*).

Ice climate (*ET*) is found mainly as a tundra subtype in rather narrow mountainous belts.

Highland climate (*H*) is considered as a separate type when any of the climates above are modified by the altitude of the region. Possible further varieties are influenced by the orography. Map showing climatic regionalisation is given in Fig. 2.30. which in the given scale can only provide general boundaries of dominant climates.

Bearing in mind the special features of each particular climate as given above, it can be concluded that:

Tropical rain climates form a belt across the equator, mainly distinguished by the absence of winter in the tropical lowlands. Frost is also absent except when the type is modified on the highlands and plateaux. The margins are determined either by diminishing annual rainfall or by decreasing temperature. The three subtypes are distinguished in accordance with seasonal distribution of rainfall and temperature. The tropical rain regions have ample rainfall for ten or more months of the year often with two peaks and two drier periods as can be seen from the climatic data for Libreville, Gabon in Tab. 2.11. Annual rainfall usually varies between 1650 and

Tab. 2.11 Climatical data for Libreville (0°25′ N, 9°25′ E, 35 m altitude)

J	F	M	A	M	J	J	A	S	O	N	D	Annual
Monthly means of daily maximum temperature (°C)												
30	31	31	31	30	29	28	28	29	30	30	31	29.8
Monthly means of daily minimum temperature (°C)												
25	24	24	24	24	22	20	20	22	22	22	23	22.7
Mean monthly rainfall (mm)												
252	252	350	345	250	20	10	20	100	280	370	200	2349

2500 mm and can be even higher when the air is more humid. Rain making disturbances are more frequent with abundant cumulus clouds producing heavy showers and thunderstorms. Most of the rain falls during the afternoon and early evening when solar radiation makes the humid air unstable. Night rains are also frequent. Rainfall periods are rather short, continue for several days and are separated by periods of clear skies. Some areas do not satisfy the criteria as given above, particularly the monthly rainfall total which can be less than 60 mm.

This type of climate is found in Malaysia, the Philippines, New Guinea, Melanesia, Polynesia, the Congo basin, the northern Gulf of Guinea, Eastern Malagasy, the Amazon basin, eastern Brazil, West Indies and the east coast of Central America. In the region of the Andes it is found up to an altitude of 1000 metres.

Monsoonal type climate is developed as a result of the difference between the temperature of the continents and oceans. A greater difference between the coolest and warmest month is typical. A monsoon forest climate can sometime be found limited to rather narrow strips along the coast where the tropical rainforest contains more deciduous trees.

Such a type of climate is found in India, Thailand, Vietnam, southwestern Sri Lanka, the western coast of Burma, Malaysia, northern Australia, and Sierra Leone. Data for Mangalore, India are given as an example in Tab. 2.12.

Tab. 2.12 Climatical data for Mangalore (15°28′ N, 75°00′ E, 24 m altitude)

J	F	M	A	M	J	J	A	S	O	N	D	Annual
Monthly means of daily maximum temperature (°*C*)												
31	30	31	32	31	29	29	28	27	28	29	31	29.7
Monthly means of daily minimum temperature (°*C*)												
20	21	23	24	24	22	22	22	22	22	22	20	21.8
Mean monthly rainfall (*mm*)												
2	2	2	40	160	920	930	560	260	195	75	2	2948

The savanna climate is under the influence of trade winds and has its dry period sharply bounded in time. The rainfall occurs in the summer, and the temperature amplitude is higher. The duration of the dry season is at least two months. This type is located on the poleward side of the tropical rain climate and is bounded on the outer sides by dry climate. It is found between the ITCZ and its rain bringing effect and subtropical anticyclones with their stable subsiding and diverging air masses on the poleward sides. With the seasonal shifting of solar energy belts and pressure and wind belts the region comes under the influence of the moving ITCZ during high sun. At the time of low sun it is under the influence of driver winds and subtropical anticyclones the result of which is a dry winter and wet summer.

The variability of diurnal temperature is considerable. Often the hottest period precedes the time of the highest sun. Thus March, April and May are usually warmer than June and July which are the rainy months north of the equator and September, October and November are usually warmer than December and January in the southern hemisphere. The annual rainfall is typically between 1000 and 1500 mm, however, an even lower amount is common. The rain comes mainly in the form of convective showers. A return of ITCZ toward the equator brings an anticyclonic drought in winter, thus the further the region is from the equator, the longer is the dry season.

The summer rain peak comes after the sun has reached its zenith. The climate is widely spread in Africa from Somalia to Zimbabwe, in the Sudan, eastern part of the Guinea Bay, the Congo headwaters, western Malagasy, parts of Brazil, the northeastern coast of Venezuela, Yucatan and Cuba. It is found also in India, Thailand and eastern Australia. Data for Harare, Zimbabwe are given as an example in Tab. 2.13.

Tab. 2.13 Climatical data for Harare (17°50′ S, 31°22′ E, 1487 m altitude)

J	F	M	A	M	J	J	A	S	O	N	D	Annual
Monthly means of daily maximum temperature (°C):												
26	24	24	25	22	20	20	22	26	27	26	25	23.9
Monthly means of daily minimum temperature (°C):												
16	16	14	13	10	7	7	9	11	15	16	16	13.8
Mean monthly rainfall (mm):												
190	230	140	30	10	5	0	5	10	20	110	120	870

Typical for the dry climate is a high fluctuation of daily temperature resulting from intensive exchange between the surface and air masses. Two subtypes – steppe (or semi-desert) climate and desert climate are recognized. The first one forms a transitional belt surrounding deserts and separating them from the tropical rain climate. According to Trewartha [38] the climate passes through the latitudinal zones and can not be defined in terms of temperature, unless rainfall is taken into account. Some sources give the limits of annual rainfall as between 250 – 500 mm others between 375 – 600 mm. The semi-deserts have a short period of heavy rain most frequently at the time of high sun. The rainfall distribution is similar to that in the wet and dry climate except that the dry season is longer and total precipitation less. Belts of steppe climate on the poleward sides of deserts are usually in fairly close proximity to the subtropical dry summer type and have nearly all their rainfall during the cool season, this being dependent on mid-latitude cyclones which tend to travel equatorwards more often in winter than in summer. In Tab. 2.14

Tab. 2.14 Climatical data for Windhoek (22°37′ S, 17°08′ E, 1665 m altitude)

J	F	M	A	M	J	J	A	S	O	N	D	Annual
Monthly means of daily maximum temperature (°*C*):												
27	27	26	25	22	20	20	22	24	27	30	31	23.4
Monthly means of daily minimum temperature (°*C*):												
17	17	15	13	10	7	7	9	11	16	17	18	14.0
Mean monthly rainfall (*mm*):												
130	90	50	20	2	2	1	1	0	40	25	40	401

data is given for the typical semi-desert climate of Windhoek Namibia. This subtype is widely distributed in southern parts of Africa, the margins of the Sahara and Australia, but is found much less frequently along the west coast of South America.

The desert or arid climate has a very hot summer and cooler winter. Extreme conditions are caused by the leeward and interior location of regions under the conditions of prevailing clear skies and low humidity. An abundance of solar energy reaches the earth by day and is rapidly lost during the night. Rainfall is always insufficient and variable. There are more years when it is below average than above. There is a high annual rainfall during very humid years because precipitation takes the form of storms, the water flows away as floods and little remains for vegetation. A humidity of 12–13 % is common for the midday hours. Dry regions often have windy days influenced by tropical anticyclones and dry trade winds. The tropical deserts receive less than 250 mm of rainfall per year; sometimes, however, even regions with rainfall of less than 375 mm are considered as deserts. Showers may fall at any time of the year but very irregularly. Low altitude desert areas have average temperature in hot months of between 30–35 °C and in cooler months between 10–15 °C. Altitude modifies the range. In some locations a temperature of 70 °C has been recorded. The diurnal range may reach 35 °C.

The modification of the dry climate, known as marine dry climate, is found along western coasts south of the equator and in Somalia. This is under the influence of the cool ocean current and is characterized by markedly lower temperature, and reduced range of monthly and daily temperatures. Rainfall is even lower than in the nearby desert due to the subtropical anticyclones and the stabilising effects of cool coastal water. Fog and low cloud cover contribute to a lowering of the potential evaporation rate.

Desert climate is found in northern and southern Africa, western South America and in central Australia. Data for Africa, Chile, is given in Tab. 2.15 as an example.

Tab. 2.15 Climatical data for Africa (18°37′ S, 70°18′ W, 5 m altitude)

J	F	M	A	M	J	J	A	S	O	N	D	Annual
Monthly means of daily maximum temperature (°C):												
23	24	22	21	20	19	18	18	18	19	20	21	20.2
Monthly means of daily minimum temperature (°C):												
19	19	19	17	16	14	13	14	15	16	16	17	16.2
Mean monthly rainfall (mm):												
1	0	0	0	0	0	0	0	0	0	0	0	1

Warm rainfall climate is not widely distributed in the tropics and the subtype with dry winter and wet summer prevails. Slighter difference between summer and winter temperature is a dominant feature. This climate is found in the southern part of the Central African Plateau, in southern Brazil, western Mexico and eastern Australia. Data for Rio de Janeiro in Tab. 2.16. can serve as an example.

Tab. 2.16 Climatical data for Rio de Janeiro (22°05′S, 43°12′ W, 61 m altitude)

J	F	M	A	M	J	J	A	S	O	N	D	Annual
Monthly means of daily maximum temperature (°C):												
26	27	26	24	23	22	22	22	22	23	25	26	24.0
Monthly means of daily minimum temperature (°C):												
23	24	23	22	20	19	18	18	19	20	21	22	20.8
Mean monthly rainfall (mm):												
125	110	140	105	75	55	40	45	65	85	100	135	1080

Ice climate is found in small localities on the slopes of the Andes in the form of the so-called ‘tundra’ climate. More frequent is the highland climate in regions where the prevailing type of climate, typical for lower altitudes, has been modified by the elevation of the region. Therefore the highland climate can not be characterized in a general way.

Exposure to the wind and sun are equally significant factors. Even isolated small valleys may have considerable variability on the windward and leeward slopes. For instance, Sesheke on the middle Zambezi has a temperature below freezing point ten days a year. An accepted decrease of the temperature is 0.65 – 0.96 °C per 100 metres of altitude.

The highlands of Mexico, Ecuador, Bolivia, Peru, Colombia, southern Africa, Ethiopia and East Africa have this type of climate. Data for Quito, Ecuador are given as an example in Tab. 2.17.

Tab. 2.17 Climatical data for Quito (0°20′ S, 78°45′ W, 2850 m altitude)

J	F	M	A	M	J	J	A	S	O	N	D	Annual
Monthly means of daily maximum temperature (°C):												
20	20	20	20	20	21	21	22	22	22	21	20	20.8
Monthly means of daily minimum temperature (°C):												
14	13	13	13	13	13	12	12	12	12	12	11	12.5
Mean monthly rainfall (mm):												
102	80	130	180	125	35	20	25	60	95	100	95	1047

2.6 REFERENCES

[1] Babu, R., Tejwani, K., G. Agarwal, M., C. Blusban, L. S., 1979. Rainfall-intensity-duration return period equations and nomographs of India. I.C.A.R. Dehra Dun 249—295.

[2] Baghrathna, V. R., Shrow, E. M., 1978. Rainfall depth-duration-frequency studies for Sri Lanka. Journal of Hydrology 37, 223—240.

[3] Bailey, M., 1970. On extreme daily rainfall in Zambia. Dept. of Meteorology Lusaka, 10 p.

[4] Balek, J., 1977. Hydrology and water resources in tropical Africa. Elsevier Amsterdam, 208 p.

[5] Balek, J., 1978. A storm rainfall pattern above the Central African Plateau. Hyd. Sc. Bull., 23, 1—3, 151—155.

[6] Barry, R. G., 1969. Precipitation, water, earth and man. Methuen, London.

[7] Cilento, R.,W., 1925. The white man in the tropics. Gov. Printer Melbourne, 168 p.

[8] Falkenmark, M., 1979. Some hydrological similarities and disimilarities between temperate and tropical countries. Nordic IHP Rep. No. 2, Hydrol. in Dev. Countries, Oslo, 17—32.

[9] Forsberg, F. R., Garnier, B. J., Krichler, A. W., 1961. Delimitation of humid tropics. Geogr. Rev. 51, 33—47.

[10] Garbell, M. A., 1947. Tropical and equatorial meteorology. Pitman, N. York, 237 p.

[11] Gregory, S., 1969. Rainfall reliability. In "Environment and Land Use in Africa", Ed.Thomas and Whittington. Methuen, London, 57—82.

[12] Griffiths, J. F., 1972. Climates of Africa. Elsevier, Amsterdam.

[13] Herbertson, A. J., 1905. The major natural regions. Geogr. Journal XXV, 300—312.
[14] Hjelmfelt, A. T., Jr., 1978. Amazon basin hydrometeorology. Proc. Am. S. Civ. Eng., JHD June 1978, 887—897.
[15] Holtz, A. C. T., 1965. Criterio de avaliação das máximas intensidades médias prováveis de precipitações intensas. Publ. No. 16 CEPHH, Universidade de Paraná, 18 p.
[16] Holtz, A. C. T., 1966. Máximos precipitações pluviais provaveis de 1 a 6 dias de duração. Pub. No. 25 CEPHH, Universidade de Paraná, 12 p.
[17] Holtz, A. C. T., 1966. Contribuição ao estudo de relação altura de precipitaçao-área-duração para chuvas intensas. Publ. No. 26, CEPHH, Universidade de Paraná, 16 p.
[18] Hudson, N. W., 1957. Erosion control research. Rhodesian Agr. Journal, 54 297—307.
[19] Huntington, E., 1924. Civilisation and climate. New Haven, Yale Univ. Press, 453 p.
[20] Jackson, I. J., 1978. Local differences in the patterns of variability of tropical rainfall. Journal of Hydrology 38, 273—287.
[21] Jackson, I. J., 1977. Climate, water and agriculture in the tropics. Longman, London, 248 p.
[22] Kendrew, W. G., 1949. Climatology. Oxford, 383 p.
[23] Köppen, W., Geiger, R., 1954. Klima der Erde. J. Perthe, Darmstadt.
[24] Lebedev, A. N., Climate of Africa. (In Russian). Gidrometeoizdat, Leningrad, 488 p.
[25] Lee, D. H. K., 1970. Climate and economic development in the tropics. Harper, New York, 176 p.
[26] Nwa, E. V., 1977. Variability and error in rainfall over a small tropical watershed. Journal of Hydrology, 34, 161—169.
[27] Pereira, H. C., 1973. Land use and water resources. Cambridge Univ. Press, 246 p.
[28] Pike, J. G., 1977. Rainfall and evaporation in Botswana. U.N.D.P. Tech. Doc. No. 1, Gaborone, 87 p.
[29] Pinto, N. L., 1972. Ábaco e considerações gerais sobre a calculo de vazão de dimensionamento de bueiros. Universidade Federal, Curitiba, 11 p.
[30] Pollock, N. C., 1968. Africa. Univ. of London Press.
[31] Raman, P. K., 1971. Rainfall climate of the project area. Hydromet. survey of lakes Victoria, Kyoga and Albert. U.N.D.P., Entebbe, 51 p.
[32] Riehl, H., 1959. Tropical meteorology. McGraw Hill, New York.
[33] Smith, A., 1970. The seasons. Penguin, Harmondsworth, 297 p.
[34] Smolík, L., Stružka, V., 1959. Engineering meteorology and climatology. (In Czech). SNTL Prague, 238 p.
[35] Souza, P. V. P., 1959. Possibilidades pluviais de Curitiba, em relação a chuvas de gande intensidade. CEPHH, Curitiba, 8 p.
[36] Tarakanov, G. G., 1980. Tropical meteorology. (In Russian). Gidrometeoizdat, Leningrad, 173 p.
[37] Temple, P. H., Rapp, A., 1972. Landslides in the Mgeta area, western Uluguru Mountains, Tanzania. Geografiska Annaler, 54 A 3—4, 151—193.
[38] Trewartha, G. T., 1943. An introduction to climate. McGraw Hill, New York, 402 p.
[39] Voeykov, A., 1883. Flüsse und Landseen als Produkte der Klimate. St. Petersburg.

3 HYDROLOGICAL CYCLE AND WATER BALANCE OF A TROPICAL BASIN

3.1 INTRODUCTION

The role of the hydrological cycle in the tropics is generally similar to the hydrological cycle in the temperate zones, however, deviations which are found in water balance calculations when compared with results from non-tropical regions result mainly from special climatic features These include the composition of the vegetational cover, the properties of soils and human activity.

The different pattern of tropical climates as discussed in the previous chapter and a comparison of the climatic factors predominating in the tropics with similar factors in the temperate zones clearly indicates their impact on the formation of the hydrological cycle.

As far as the vegetation is concerned, biologists regard as a characteristic feature of tropical plants the multiplicity of species and subspecies with relatively few individual plants in any one species. A mixed population with varying demands can exist where the physical and thus also the hydrological situation encourages reproduction and survival but the storage of any one requirement is limited. Of the large number of species and varieties of plants that may be found in tropical vegetation many are useful to man. Thus the great potential of tropical vegetation results in the increasing effect of human activity on the hydrological regime. In contrast to the temperate zones man's attention in the tropics has been focused on regions which from one point of view are more economically exploited on a large scale. Such activity has a negative effect on the natural hydrological cycle of vast regions.

Unfortunately, in the future it will be difficult to change such negative effect. Abundant rainfall on the one hand together with irregular or very limited rain on the other, the unpronounced seasonal variability of daylight, the shortage of some nutrients and diminished photoperiodicity are the phenomena which make it difficult to introduce to tropical regions many plants from temperate zones.

As can be expected, tropical plants are rather poor in nitrogenoeous constituents which are contained in tropical soils to a rather limited extent. Actually, the exhaustion and restoration of the soil is a delicately balanced process in the tropics. The traditional practice of burning the natural forest together with natural forest fires, combined with growing crops for a few years until the soil becomes exhausted and the subsequent abandonment of the area can easily be destructive to the soil.

It has been fully successful only in certain places where the practice of tree planting has been systematically introduced.

Strip-farming in which only small areas are cleared and planted while the surrounding forest remains intact, is perhaps more economical from the point of view of water management. However, there are several aspects that have to be considered, particularly the exposure of the strips to erosion and the leaching out of soluble nutrients.

On the other hand many tropical plants tend to be rich in carbohydrate which is synthetised from carbon dioxide in the air and water.

Thus natural and induced changes of tropical vegetation produce serious problems in the water balance calculations of many tropical basins. For instance, the root system is normally adapted to the groundwater level and vice versa and any change in it will result in a change in the capillary rise, soil moisture storage, groundwater storage and baseflow formation. A decrease of interception by introducing plants with reduced leaf areas leads to an increase in surface runoff and erosion under intensive tropical rainfall.

It is not only the direct activity of man that produces such effects. For instance, in Luangwa Valley Park along the Zambian and Malawian borders increasing herds of elephants destroyed the natural vegetation cover to such an extent that a rapid growth in the intensity of the erosional processes could be traced. Similar effects were reported from Zimbabwe by Savory [61]. Such experiences clearly indicate the delicate nature of the stability of the water balance of tropical watersheds.

The existence of luxuriant vegetation in many parts of the tropics and the rapid growth of some species have resulted in the belief that food in the tropics can be provided without much effort. However, many plants from the temperate zone simply cannot grow in the tropics because they require a certain amount of energy input and a certain ratio between day and night. A net photosynthesis is given as the difference between photosynthesis and respiration. Photosynthesis is determined as

$$CO_2 \rightarrow CH_2O + O_2$$

which indicates that daylight and cloud cover are significant stimulants. Respiration is written as

$$CH_2O + O_2 \rightarrow CO_2 + H_2O$$

and is a continuous process in the majority of plants. In general it increases with higher temperature and thus tropical climatic conditions remain favourable only for certain types of species.

3.2 THE ROLE OF VEGETATION IN THE HYDROLOGICAL CYCLE

The role of tropical vegetation in the tropical hydrological cycle has been studied for a long time and different opinions can be found in the conclusions reached. In general the stabilisation of the hydrological regime is of benefit to the natural vegetation, particularly to the tropical forest. The reduction of the flood regime and an increase in the low flow during the dry periods are regarded as stabilising factors.

A sceptical approach stresses the negative role of vegetation in that it increases the rate of transpiration through the well developed leaf area in a mature forest as compared with poor and less dense vegetation. Experiences from temperate regions indicate no significant difference between an afforested basin and a basin covered by grass, providing the soil is permanently well saturated. Another set of experiments proved that there is a lower total annual outflow from the afforested regions when compared with deforested regions. The latter, however, have a higher amplitude of annual runoff. It is not an easy task to quantify these conclusions because in the tropics few experiments have been carried out and due to a different rainfall pattern and increased evapotranspiration, results from outside the tropics cannot be directly applied. In the Luano catchments in Zambia much higher flood peaks have been recorded on the deforested catchments cleared of the Miombo forest, compared with the uncleared areas, while during the period of observation when all catchments had been forested their behaviour was very similar.

It is hard to apply the results from small areas, even if performed in the tropics, to extensive regions of the tropical forest. A different approach needs to be followed, based on the observation of gauging stations on large streams and data from hydrometeorological network of afforested and deforested basins. However, another problem immediately arises: does the forest control the rainfall regime of the region or is the rainfall the dominant factor? Bernard [9] after studying the Congolese tropical rainforest, then covering about 1 million square kilometres, found no evidence of the rainfall being influenced by the forest, apart from local instability resulting from increased heat reflection from cleared areas.

In the vicinity of the Kariba dam an increase in annual rainfall of 5 – 10% was traceable after the establishment of the lake. In contrast, Taylor [66] assumes that the air above tropical forest enriched by the transpired water can develop orographical rainfall if it is forced to ascend. As early as in 1887 Blanford [10] found that after afforestation, the rainfall above the area increased by 13%, however, these results from India were later attributed to the low density of the observing network. Nevertheless, a similar effect was reported in 1905 by Lafosse from southern Egypt.

While the effect of the forest on the rainfall regime remains in dispute, there is much less doubt that deforestation has an impact on the desertification of the region through the degradation of the soil. Intensive cropping of sugar cane and coffee in Brazil, peanuts in Senegal and cocoa in west Africa and Brazil resulted in soil degradation after several years of profitable production. In addition, the destruction of the soil structure and wind and water erosion contribute toward the increased desertification of the soils. At this point, the conclusions of Ayyad [2] should be recalled on the desertification of parts of Egypt by overgrazing, over-cultivation and felling of timber in the region which was once "the granary and vineyard of Rome". The region has been turned into a desert because plant cover has been destroyed beyond the minimum coverage needed for the protection of the soil against wind erosion. At present the mean annual rainfall for that region is between 120 and 180 mm per year. This is not enough to support vineyards and wheat without extensive irrigation. This only indicates that some changes of the rainfall pattern came about either because of changes in the vegetational pattern or by some long-term climatic changes. A similar change was reported by Hollis [30] from Tunisia, where historical settlements along Lake Ichkeul indicate that the lake level was once much higher than it is today.

Another negative impact is that of bush fires either controled or uncontrolled. Changes in the hydrological regime can be even more drastic than those created by slow continuous felling when open stands are covered at least by some vegetation.

According to various sources such a sudden change can increase the runoff volume tenfold and soil losses 100–3000 fold. Such a type of effect has been described in the coastal parts of Congo near Point Noir and on sandy deforested soils in Minas Gerais and Sao Paulo in Brazil.

Not only the quantity but the quality of water is influenced by the clearing of forests. Reduced use of water by forest transpiration leads to an increase in the groundwater level. In some regions the springs become salty. Such a drastic change was reported from Australia when deep-rooted Eucalyptus species were replaced by the rotation of wheat and short season grasses and clovers.

In contrast to this, in South Africa forestry plantation came under attack from farmers who claimed that forest plantation had a negative effect on water supply. At one time it was prohibited to plant trees along river banks and within twenty metres of springs. Nevertheless, such an approach can be disputed. It is quite possible that in warm latitudes where rainfall occurs in the form of violent storms and in concentrated rainy seasons separated by hot weather, the hydrological balance is more sensitive to disturbance, because bare soil is unable to absorb storm rainfall and the increased overland flow results in reduced baseflow.

The problem of cutting down tropical rainforest, or as it is called by Ewel [19] "forest mining", has become a matter of serious concern to environmentalists. It follows from numerous reports that the region of the Amazon tropical rainforest appears to be seriously endangered. Ewel feels that the renewal of the tropical rainforest should be encouraged in order to avoid extensive soil erosion, loss of nutrients and the reduction of the microorganism population and habitat. A typical environmentalistic point of view is represented by Poore [54] who sees the role of the tropical forest as being similar to that of the forest in temperate regions, including the recreational effect and the provision of a subsistence basis for the life of the indigenous people.

It has been estimated by Nigh and Nations [48] that about 250 000 km^2 of tropical forest is vanishing annually. Rantjisinh [57] studied forest destruction in Asia and the South Pacific. He found that in this region about five million hectares are lost annually. In Thailand, for instance, the proportion of the area of forests in the country decreased from 58.3 to 33% within twenty-five years. According to Rantjisinh, improper land use can be blamed for increased effects of flooding in India, Pakistan and West Africa. Aiken found the situation critical in Malaysia. Mooney [46] believes that the resettlement of hundreds of thousands of families along the arteries of the Transamazonica Highway together with the replacement of the forest by roads, farms and ranches, threatens the rainforest ecosystem. Also in the semi-deciduous forest of Central Africa an increase of surface runoff from areas with new unpaved road systems was observed.

Gwyne [27] stated that the belt of greenery around the equator remained unchanged for 50 million years and that the increasing rate of its present destruction must have long-term and large-scale effects. This opinion is shared by Brunig [13] who concludes that unfortunately such an effect cannot be evidenced on model catchments and on small scale. Nevertheless, Potter et al. [55] have examined possible climatic changes by a two dimensional atmospheric model based on the equations used by conservationists and have reached the conclusion that besides other consequences global cooling and reduced precipitation can be expected.

Medina et al. [45] attempted to make a complex study of the natural ecosystems in the region of the Venezuelan Amazon basin, including the perturbation and the cutting down and burning of the woods on experimental plots.

A so-far controlled, extensive experiment is reported from the semi-arid region of the Argentine part of Gran Chaco, where in a pilot area of 53 thousand hectares unproductive bush and forest were cut down at the rate of 4 to 5 thousand hectares per year and transformed into grazing land with a high production of beef. It is planned to similarly transform twenty million hectares in Argentina and forty to fifty million hectares in Bolivia and Paraguay. In the region, however, there was no water supply, nor a type of climate which could possibly be destroyed or at least damaged.

Nevertheless, the forest in arid lands also has a certain hydrological role and for this reason Debano claims that it should be protected [17]. Also Went [72] supports the protection of the sparse vegetation in arid regions. Such a type of vegetation is not at present used commercially but for shelter and as a fuel which promotes aridity.

Large-scale deforestation can be traced back in some regions to the 18th century according to reports by Veblen [71] from the Guatemalan highlands. It remains difficult, however, to evaluate the consequences of deforestation in numerical terms with the exception of the cost of the timber which has been felled. A proper solution has to be sought somewhere between two possible extremes, indiscriminate forest utilisation on one hand and total conservation on the other. Problems of an impact analysis approach as applied to forestry management were analysed by Teller [67]. An example of such an analysis is given in the form of the simple flow chart in Fig. 3.1. Assuming that in the distant future tree removal will continue

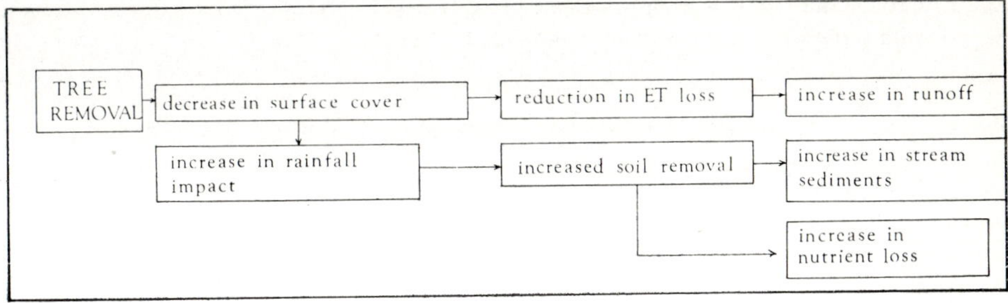

Fig. 3.1 Impact analysis approach as applied to forest hydrology, according to Teller

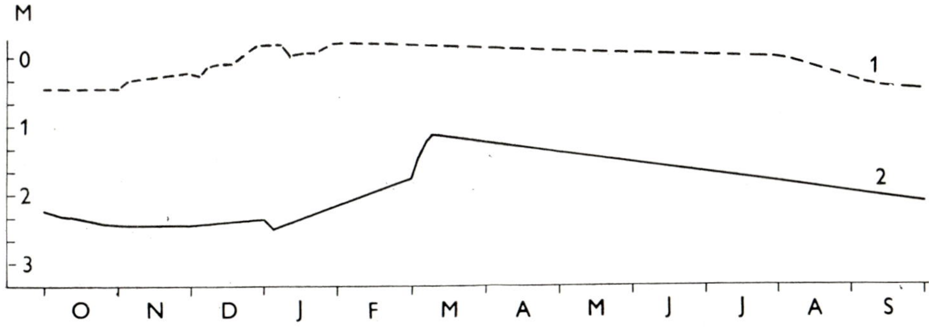

Fig. 3.2 Groundwater level fluctuation under Miombo forest (1) and under grassland (2)

at an even greater rate on areas of extensive size, it can be expected that effects of tropical rainfall will increase for two reasons. First, because of the high rainfall rates of tropical storms as compared with rainfall in temperate regions and second because many afforested watersheds in temperate regions are adapted to reduced vegetational cover for some part of a year. Thus any conclusions and the application of the formulae, particularly on soil removal and on erosional processes need to be re-examined. Many tropical forest species gain water from groundwater resources through the zone of capillary rise. Not only surface runoff, but baseflow will also very likely increase in many tropical watersheds. Nutrient loss will be highest at the initial stage of the process due to the intensive erosion of the upper soil layer.

A comparison of the groundwater level fluctuation under Miombo forest at the Zaire-Zambian borders and under the grassland in the same area can be seen in Fig. 3.2. In the grassland the morphological conditions for the formation of groundwater storage are very similar to those in the forest, however, the transpiration rate is much lower and the groundwater level more stable than that below forests. Also the impact of different vegetational cover can be seen in Tab. 3.1. The

Tab. 3.1 Groundwater outflow from woodland and grassland in Miombo forest (mm) as compared with surface runoff, total runoff and rainfall (mm)

Month	Woodland	Grassland	Surface	Total	Rainfall
October	0.79	0.00	0.04	0.83	59
November	0.51	0.00	0.07	0.58	83
December	0.86	0.81	33.42	35.09	427
January	3.40	4.80	63.31	71.51	341
February	6.30	4.34	59.10	69.74	264
March	8.13	4.80	17.73	30.66	23
April	5.82	4.24	2.94	13.00	38
May	4.32	2.92	0.42	7.66	0
June	3.10	1.25	0.00	4.35	0
July	2.29	0.38	0.00	2.67	0
August	1.23	0.00	0.00	0.00	0
September	0.29	0.00	0.00	0.00	8
Total	37.04	23.54	177.03	237.61	1243

difference between the runoff from afforested part of the watershed indicates the influence of the vegetational cover on the hydrological regime in Miombo forest area. Although that part of the watershed which was uncovered, or in other words, formed by roads, tracks, bare soil etc., was rather insignificant as compared with the total size of the watershed, the surface runoff formed a significant part of the total runoff. An intermittent regime of annual flow has been typical in this part of the tropics for some years and depends on the groundwater storage accumulated during the wet season. In the grassland an additional factor is the date of the last rain during the wet season.

It is rather surprising to find such a high surface runoff as indicated in the table. This is developed mainly through the overstored groundwater aquifer and this feature can serve as an indicator of the difference between the influence of forest and grassland on the hydrological regime. A similar and even more pronounced effect was found in the Ngono basin west of Lake Victoria in Uganda. The direct surface runoff from the tropical forest is negligible regardless of high rainfall intensity and the high soil moisture content. A similar conclusion was published by Dagg and Pratt [16], based on the observation of the Sambret and Lagan watersheds in Kenya. Here, less than 10% of the total annual runoff was formed by surface runoff.

Pitman [53] studied the impact of afforestation of the southern rims of the tropics in South Africa and found that afforestation of the grassland increased the interception, evapotranspiration and also the abstraction of soil moisture. He developed a formula for calculating the reduced rate of the mean annual runoff parallel to the increasing rate of afforestation:

$$Q = 34.4 - 0.867(x - x_i)$$

valid for an afforestation rate of 3.6 km^2 year^{-1} and a basin of 200 km^2. In the formula, Q is mean annual runoff (10^6 m^3 year^{-1}), x_i is the calendar nunber of the year when the afforestation was initiated and x is the calendar number of the year for which the mean annual runoff is calculated. Such a type of formula is not directly applicable to any region and can be regarded only as an indication of a possible approach to the solution of such a problem.

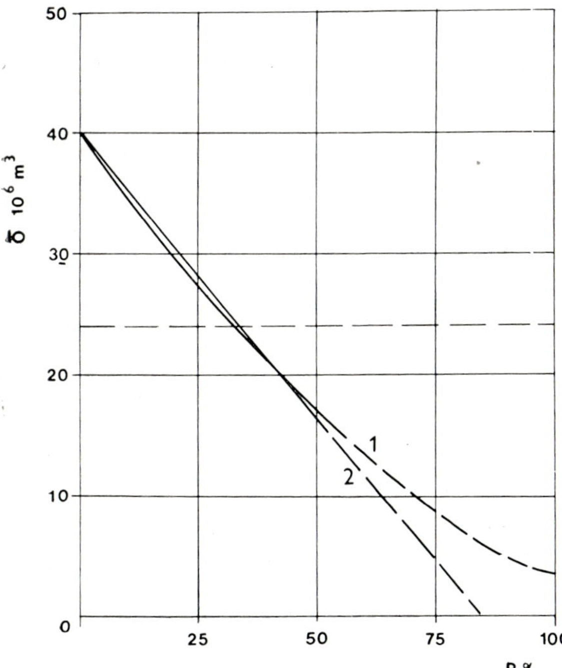

Fig. 3.3 Decrease of annual runoff depending on the size of the afforested area, after Pitman. 1, actual relationship; 2, simulated relationship

An interesting graph indicating the decrease of the annual runoff depending on the size of the afforested area can be seen in Fig. 3.3. and was developed by Pitman [52].

A comparison of the percentages of different types of land use is given in Tab. 3.2.

Tab. 3.2 Land use on the continents (hectares 10^{-3})

Continent	Afforested	Agriculture	Other	total	% of afforested area
Asia	519 960	734 140	1 463 470	2 717 570	19
Africa	752 800	549 720	1 674 890	2 977 410	25
South America	956 660	403 280	414 900	1 774 840	54
Central America	74 030	107 920	89 980	271 930	27
Australia, Oceania	95 870	44 670	713 860	758 530	27

The role of clearcutting on the hydrological regime performed on a small scale within an otherwise undisturbed environment can result in different effects that can be seen on a large scale. However, Whyte [74] found that the disappearance of woody plants even on a small scale and/or the reduction of their average height, worsens the micro-climate because the action of the wind and solar radiation on plants and soils becomes more marked, temperature variations become greater on and below the soil surface and trees and shrubs are no longer reproduced. This can create dangerous situations especially when a desert fringe type of vegetation, acting as a protective barrier between the desert and sub-humid regions, is cleared.

Several attempts to summarize the effects of the tropical forest on the hydrological regime have been made and emphasis has ben placed on different aspects. As early as 1949 Wicht [75] summarized his observations of regions of South Africa and found that:

a) forest will use more water than grasses;

b) the amount of water consumed by forests depends primarily on the amount of water available in the soil;

c) swamps will dry out if trees are planted in them;

d) the removal of the vegetation will increase the discharge.

Similarly, Matějka [43] concluded for the tropics:

a) forests hold the rainfall and protect it from flowing away. Water is evaporated more extensively from forests than from open stands and evaporated water contributes to the enrichment of newly forming rainfall;

b) the cooling effect of the crowns of trees and increased evapotranspiration accumulate cool air above the forest and form favourable conditions for the additional condensation of water vapour;

c) in contrast to grasslands the atmospheric pressure gradient of the forest decreases from upper to lower layers and thus may contribute to the formation of rainfall;

d) similarly the temperature decreases with height above the forest and thus also contributes to the formation of rainfall;

e) the ionizing effect of the air contributes to water vapour condensation.

Balek [5] concluded for the Miombo forest in wet and dry regions that:

a) the variability of the rainfall pattern is not influenced by the forest;

b) the most significant component of the water balance of tropical afforested watersheds is evapotranspiration;

c) the root system is adapted so that the roots can tap water through the zone of capillary rise. While the upper root systems tap a similar amount of water year by year, the lower root systems tap an additional amount depending on the depth of the groundwater level;

d) forest interception forms a significant part of evapotranspiration during the wet season. The evapotranspiration rate during and after rainfall may exceed the potential evaporation for the same period.

When generalizing similar types of experimental results, one has to bear in mind the wide variety of forest types. Some foresters even suggest the use of the term 'forest formation' instead of forest, particularly where different types of forests are mixed in keeping with changing orography or when one type of forest is continuously changing into another. For low latitudes of the tropics rainforest is most typical, this being transformed as the latitude rises to mixed deciduous forest. With the increasing length of the dry period, semi-deciduous forest is replaced by deciduous forest. At the edges of the tropics the number of evergreens steadily increases, some of these, however, are induced – for instance various types of pines in South Africa.

In the low precipitation belts succulents of various types are found and in a great part of wet and dry tropics Miombo forest and baobabs.

Xerophytic shrub consisting of various drought-resistant and low growing plants are typical for semi-desert and desert regions. In Argentina formations of low dwarf trees are called 'monte'. Here, the vegetation along the rivers consists of different species called tunnel-forest or 'galería' with dense foliage above the stream channels.

On the cool side of the deserts a scrub evergreen forest and brush of a broadleaf type are found while on the warm side broadleaf deciduous types are dominant. The term 'wooded savanna' is sometimes used for transfer belts between dry savanna and semi-deserts.

A difference is made between the tropical rainforest, in South America also called ‘selva’, and the jungle which is a term used for the forest with spots of tangled underbrush on its shaded floor. Such a type of underbrush can also be found in the rainforest and semi-deciduous forest.

An effect of the forest cover as compared with the savanna can be seen in Tab. 3.3.

Tab. 3.3 Evapotranspiration (mm) from forest and savanna of Central African Plateau, compared with potential evaporation and rainfall

Month	Potential evaporation	Annual rainfall	Evapotranspiration forest	Evapotranspiration savanna
October	190.97	95	66.98	29.13
November	171.11	169	152.73	124.00
December	145.98	244	145.98	65.86
January	146.42	358	146.42	99.64
February	152.87	161	162.95	71.25
March	161.63	241	197.69	55.30
April	150.93	7	127.89	26.59
May	130.22	0	87.45	10.34
June	101.38	0	63.36	9.02
July	114.75	0	61.38	6.64
August	136.40	0	55.04	5.56
September	159.87	2	46.71	5.33
Total	1762.53	1277	1314.58	508.46

Vast regions of the Asian tropics are covered with deciduous monsoon forest of the wet and dry type. Also mangrove forests in the coastal regions of river mouths form a distinctive unit, like forests in the marshlands.

Under the influence of warm trade winds cooled and condensed air moisture contributes to the formation of the fog forest, also known as weeping forest on the eastern slopes of the Andes. In Paraguay there is a zone of marshy lowland characterized as wet savanna forest and formed by tall grasses and distinctive belts of trees.

Beard [7] recognized three transitional types between evergreen and deciduous tropical forest:

a) semi-evergreen seasonal forest,

b) deciduous seasonal forest,

c) thorn woodland.

The first type is common, according to the author, in the Brazilian cerrado and West Africa and as a wetter monsoon forest in Asia. Deciduous forest in Africa is equivalent to the Miombo type and in Asia to the drier monsoon forest. Thorn woodland is typical of a great part of the American tropics, for instance in northeastern Brazil.

As will be discussed later, the forest type is a dominant factor in the hydro-ecological regionalization of the tropics.

There are also many varieties of native grassland of tropical semi-arid and arid regions, called steppes, prairies, savannas or pamps. The designation of different types of grassland is far from uniform. For instance, the so-called pantanal type, which covers an area of 250 000 km^2 in southwestern Brazil, is typical of regions which under wet summer conditions and about 1200 mm of rainfall are temporarily changed into extensive swamps with islands of drier and partially afforested hills. In Paraguay there are extensive areas of marshy lowlands with vegetation designated as wet savanna. In cooler regions steppe land is formed by short grasses covering the ground with a sod. Where the climate becomes more humid tall prairy grass covers the land in the form known in Argentina as humid pampa, which differs from North American prairies by the bunchy formation of grasses. Vast areas of Latin America, especially the interior of Brazil are characterized by tropical savanna.

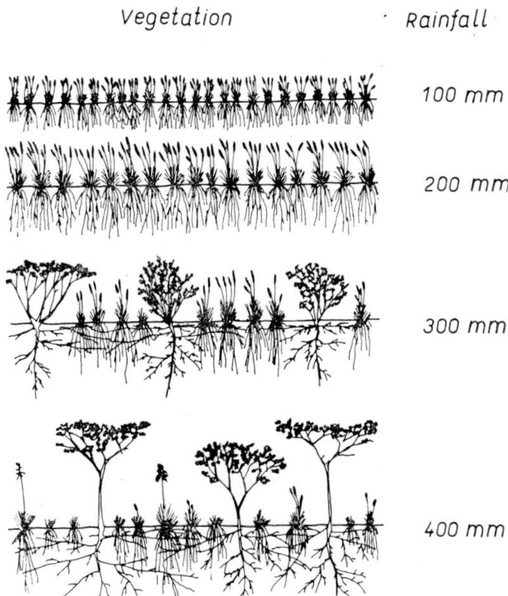

Fig. 3.4 Scheme of transition from grassland to savanna and tropical dry forest with increasing rainfall, according to Walter.

While wet savanna is found in areas subject to frequent inundation and consists of tall rank grasses with an absence of trees, dry savanna is found in the drained areas and is subject to frequent or regular drought. The latter type is mixed with scattered shrubs and transformed into the tropical shrub forest.

It is estimated that about 30% of the tropics is covered by various types of grassland.

Walter [73] differentiated between pure grassland and savanna. His graph, which has been developed for deep sandy soils (Fig. 3.4) indicates how the vegetational composition changes with increased annual rainfall. With 100 mm of rainfall per year only metre of the soil layer becomes moist and thus only grasses with shallow roots survive. Only when 300 mm of rainfall falls in a year, are conditions favourable for shrub and tree growth and tree savanna originates. There is another limit, about 500 – 600 mm, at which trees form a closed crown canopy and grass becomes merely a tolerated species. Then the tree savanna is transformed into some form of tropical forest.

Such a balance is found very rarely under natural conditions. In addition to rainfall, soil structure and groundwater storage play an equally important role. For example in Australia in a region with an annual precipitation only slightly above 250 mm there exists 'Triodia steppe' with strips of acacia called 'mulga'.

Sometimes in regions with rainfall typical of a certain type of vegetation the groundwater table is not formed at a depth available to the roots and a different type prevails. Savanna of anthropogenic origin may replace the timber burned for charcoal or follow unsuccessful attempts to cultivate the soil. In extreme cases man-made desert may replace savanna as a result of overgrazing.

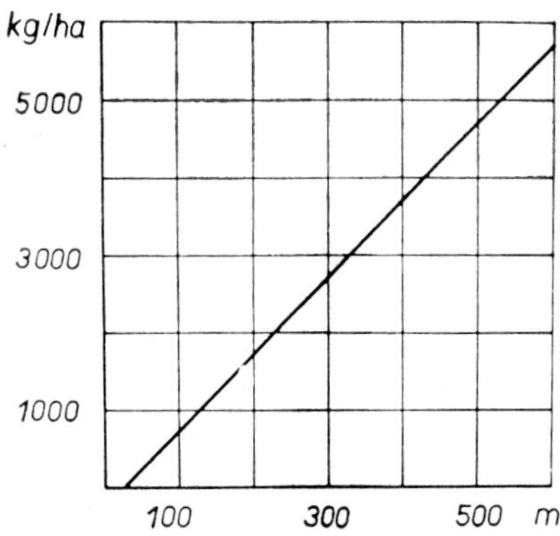

Fig. 3.5 Productivity of grassland in relation to yearly rainfall, according to Walter

During the dry season grass can survive with only a small amount of water stored in the soil and as concluded by Walter, the transpiration surface of the grass decreases in proportion to the annual rainfall in such a way that the water supply of a unit transpiring surface is of the same order under extremely arid or more humid conditions.

This rule is geographically explained in the linear relationship between the productivity of grassland and the average annual rainfall (Fig. 3.5).

In a given region the actual effect of grassland on the hydrological regime is seen after the grassland has been replaced by forest or crops or after the grassland has been overgrazed. Also the change of grass type to another type with a different root depth may have significant results.

Tab. 3.4 shows evapotranspiration calculated month by month for the Miombo forest and dry grassland. The evapotranspiration/potential evaporation ratio varies considerably month by month for both vegetational covers. The highest value of evapotranspiration from grass came in the second month of the rainy season which can be accounted for by the growth of new grass. During the dry season the amount of evapotranspiration was no more than 14 % of forest evapotranspiration and in a year grass transpired only 39 % of the amount of evapotranspiration observed in the neighbouring forest. During a year grass transpired only 29 % of the potential evaporation, while forest transpired 75 % of it. During the wet period the amount of actual evapotranspiration exceeded the potential evaporation. Even if these values vary from year to year, they clearly indicate the influence of the hydrological regime, which can be expected after changing the vegetational cover. It has been estimated that the removal of grass will contribute to an increase of the surface runoff by 370 %.

Tab. 3.4 E_t/E_0 ratio for some tropical crops (Nieuwolt)

Crop	Months after planting						
	1	2	3	4	5	6	7
Groundnuts	.45	.80	.90	.90	.90		
Bananas	.40	.50	.60	.70	.80	.90	1.0
Sugar cane	.30	.50	.70	.90	1.00		
Maize	.55	1.10	1.20	1.20	1.20		
Sorghum	.80	1.20	1.20	1.20			
Alfaalfa	.55	1.00	1.00	.60	1.60	1.00	
Tea	.00	.50	.85				
Eucalyptus	.40	.55	.70	.80	.95	1.05	1.30

Savanna lands cover approximately 13 % of the globe's surface. While in Central America savanna is not too extensive, in South America vast areas of so-called 'llanos' exist in the Orinoco basin and campos in central and southwestern Brazil. In Africa savannas extend across the continent south of the Sahara and on the Central African plateau. In Australia they are found to the south of the tropical rainforest. Normally savannas do not border directly on the rainforest, which is transformed in keeping with the length of the dry season, to seasonal forest and woodland and woodland changes to savanna. The replacement of savanna by an agricultural area is regarded as being risky. Particularly when scrub savanna is to be replaced under semi-arid conditions and with limited possibility of irrigation. More suitable is a transfer to cultivated grassland utilised for ranching. Such a project was reported from Argentina where the introduction of sweet clover, Rhodes grass and sorghum was successful.

Many of the problems of the past and still more of the present influences of man on the hydrological regime in the tropics is relevant to desertification. While many of the semi-arid to arid areas have remained more or less stable within our era, being influenced only by the climate, the situation is getting worse in vast areas under the additional influence of man. At present the problem of desertification is among the most serious in world water resources management. In accordance with the United Nations Development Program "... desertification is the impoverishment of arid, semi-arid and some sub-humid ecosystems by the combined impact of man's activities and drought. It is the process of changes in the ecosystems that can be measured by reduced productivity of desirable plants, alternatives in the biomass and the diversity of the micro- and macro-fauna and flora, accelerated soil deterioration and increased hazards for human occupancy".

It should be pointed out that desertification does not refer to totally devastated land only. It is recognized in various types: light, moderate, severe and very severe. The first type is characterized by minor effects, the second by plant-cover deterioration, the occurrence of small dunes and gullies indicating the impact of wind and water erosion. Soil salinity reduces the crop yield by no more than 50 %. In the third type undesirable shrubs and forbs replace more desirable grasses, land is denuded by sheet wind and water erosion and large gullies are formed. Soil salinity developed by the drainage and leaching reduces the crop yield by more than 50%. Large and shifting dunes are formed and large gullies and salt crusts are developed. The last stage is considered as economically irreversible and the land lost to desert forever. Remarkably, deserts with less than 100 mm of annual rainfall belong to the first category, because these regions have very little which can be destroyed and the changes rather than the present stage are used as criteria.

For an assessment of desert conditions, Budyko's aridity index, written in various forms is used. In a simple form it can be written as

$$A = \frac{E_r}{P_A}$$

where A is the aridity index, E_r is the amount of water which can be evaporated owing to local radiation, and P_A is the annual precipitation which is actually available for evaporation. While for the whole earth this index is estimated at 1.0 – 1.2, for extremely arid conditions it can exceed 200.

While it may appear that there is very little to be said about the desert vegetation

reverse is true. The biological and the hydrological regimes have perhaps a closer relationship in the desert and semi-desert regions than elsewhere. The main source of water in deserts is rainfall and dew. The latter can be highly significant in coastal deserts and is utilised by many desert plants. Balancing such types of regions on should bear in mind that the rainfall water enriching the hydrological regimes in these regions, can be of exotic origin flowing into the desert region from outside sources. Such an exogenous supply can be found especially in the margins of the deserts where they border on more humid highlands and the groundwater storage is enriched from there. Another source can be the recharge from the river crossing the desert. Two types of rivers can be found, one where the river crossing the desert remains perennial as in the case of the Nile, and the other when the perennial regime of the river is transformed into an intermittent or ephemeral regime as is the case with some rivers flowing from the Ethiopian highlands. Another type of source can be historically exogenous when groundwater resources have been formed a long time ago under different climatic conditions. In principle the area benefitting from such sources is usually limited to rather narrow strips along the river or in the vicinity of humid regions.

The Sahara region can serve as an example. Once it was like the present Mediterranean coast of northern Africa. Much evidence that the rainfall was once abundant is provided by the fossil fauna and flora. A moderate climate existed there about 10 000 – 6000 B.C. and the vegetation was very similar to that of Atlas today (Furon [20]). The temperature rise continued until 3000 B.C. and the climate worsened up to 2700 B.C. Mediterranean flora was gradually replaced by xerophylic flora and the lakes and pools which once accomodated fish, hippopotamuses, crocodiles and turtles, disappeared. The dryness gradually transformed the Sahara into steppe and then into desert. Hunters and fishermen were replaced by pastoral nomads. It is highly probable that real desert conditions were not formed earlier than the first century A.D. In this case, at least the centre of the Sahara is a type of desert produced by the climate alone, however, the stage of desertification at the margins which can be witnessed in our time is at least partially accounted for by human activity.

Surface flow in true deserts is very sporadic and deviations from the average can be higher than the average itself. As Smith [63] states the figures for mean annual rainfall widely used elsewhere as climatic characteristics, are often meaningless in many parts of the tropics. Individual rainstorms have to produce a certain amount of water to become effective. This corresponds with the amount of potential evaporation because below such a limit the rainstorms become ineffective to plant life. On the other hand water flows away as a torrential flood from rainstorms above the upper limit, and again, vegetation can gain little benefit from it. In wadi stream beds, dry for several years, violent storms can cause dangerous floods which can occur unexpectedly at any time of a year.

Water infiltrating into the soil or rocks fills the channel bottom where topography permits and forms watering places. The water stored in shallow layers is subject to actual evaporation and to the orographic pattern. Of course, there is no rule for the infiltration rate in desert regions and the permeability and porosity of desert channel beds and surface layers is highly variable.

Because the drainage system is rather undeveloped, with the exception of the beds formed before the present stage of aridity, the water flowing through the channels may reach various types of pans without any drainage outlet and then an intermittent or ephemeral type of standing water or swamps is formed. Perennial types of swamps and lakes are formed very rarely in arid regions and only under particularly favourable conditions. Overland type of flow under such conditions results in the transport of dissolved mineral matter and increasing salinity of the transported water. Due to the evaporation in closed pans saline incrustation is formed on the bottom of the pans. Therefore, intermittent and perennial lakes tend to be highly saline.

Bagnold [3] observed that in sandy desert regions surface water penetrates quickly into the soil, however, owing to capillary tension it can sink only to a certain depth which is assumed to be no more than eight times the immediate precipitation. Thus at depths of 20 – 30 cm a moist unsaturated zone is formed while the sand above and below is dry.

For the same reason groundwater storage can only very rarely be enriched and occasionally perched water level bodies may be found at local impervious depths. The water stored in the bottom layers of intermittent and ephemeral streams, which may form a locally significant source of water, is also regarded as a perched water table. Water in the saturated zone at rather extensive depths frequently fails to reach the surface naturally.

Desert vegetation is often divided into xerophytes, phreatophytes and halophytes. Hydrologically xerophytes are adapted to the economical consumption of water, functioning in a dormant stage when drought is prolonged. Phreatophytes are deep rooted and capable of consuming water from the capillary fringe, and perhaps directly from the groundwater table. Phreatophytes a root length of twenty metres or more can serve as indicators of the presence of groundwater. Halophytes are salt-resistant types of plants found where the water is highly mineralised and they can belong to either of the two previous groups.

As stated by Logan [41], deserts are not necessarily characterized by great heat nor are they necessarily vast areas of shifting sand dunes. Aridity is the basic factor or in other words lack of moisture. In extremely porous soils water is allowed to percolate so rapidly that little is left for the use of plants and thus an edaphic desert is created. For instance coastal sand dunes in areas with a rather high amount often display desert characteristics. A typical example is the Kalahari desert which is shown on the maps as occupying all of Botswana and the eastern part of Namibia.

However, much of central and northern Botswana and the eastern part of Namibia absorb water from an annual rainfall of 200 – 500 mm and the region can be considered as of the edaphic type, while only the southwestern part of Botswana and parts of Cape Province in South Africa are real climatic deserts.

The transpirational process of desert plants has been studied since the beginning of the century. As early as 1907 Livingston [40] found that water loss from three species of cacti was at its maximum during darkness and at its minimum during the day. It was generalized by Gindel [22] that the higher the value of climatic factors, including temperature and sun energy, the lower the transpiration rate of desert species. This mechanism is regulated by the closing and opening of the stomata, which are able to close under very intensive light. Thus water retention in desert plants is affected by their structural and physiological characteristics, a fact not taken into consideration in the transpirational formulae produced by hydrologists. Some plants are even able to shed their leaves as the soil moisture decreases, others can drop their leaves as the soil moisture decreases and gain new ones immediately after the next rainfall. Up to five annual crops of leaves have been observed on *Fouquieria splendens* in Australia. Therefore future research in the field of desert hydrology should be concerned with the relationship between the area of the transpiration surface, absorbing surface and climatic factors, including precipitation. A conclusion reached by Walter [73] can be cited: "... the vegetational cover is held to be proportional to the precipitation so that per unit of transpiring surface, plants in these habitats receive the same quantity of water".

The Atacama desert in Chile can be cited as an example of a true desert. It is one of the few spots on the earth where no rain has ever been recorded and where the surface remains barren over vast areas. Only Río Loa flows across it. At one time dry basins called bolsons, which separate the coastal range from the Andes, and which are some 80 kilometres in width and at elevation about 700 metres, were filled with water but now they are invaded by alluvial fans. With increasing aridity the water disappeared and salts are deposited there in the form of caliche composed of sodium chloride, sodium nitrate and some other salts.

At Iquique there was no rain between 1899 and 1919 and in the following six years the total was less than 30 mm. At Calama no rain has ever been recorded. The first has a mean humidity of 81 %, the second 48 %. Many valleys descending from the western Andes have surface water only at irregular intervals and streams of the wadi type. However, a sheet of water is continuously seeping down through the grooves of Andean ravines and the alluvial fans beyond the mountains act as sponges permitting the water to sink beyond the reach of ordinary wells, with the exception of a few west bolsons where the water comes closer to the surface and within reach of the plant roots.

3.3 ATMOSPHERE-PLANT-SOIL WATER SYSTEM

In the water balance calculations of tropical watersheds evapotranspiration is the most significant phenomenon. From a biological point of view the movement of water through a plant is described as a process which depends on its chemical potential, or in other words, on the water deficiency in a plant. Molecules of water tend to move from the site at which they have a higher energy content to the points where the energy content is lower. Obviously in the tropics and particularly in their semi-arid and arid parts the gradient of the energy content is more pronounced because of the increased influence of solar radiation.

According to Fick's law the difusion rate of a substance depends on the gradient of the water potential, the membrane cross-section, the temperature and the concentration.

The water potential corresponds to the relative chemical potential ψ so that:

$$\psi = \mu - \mu_0, \qquad \text{J mole}^{-1}$$

where μ is the chemical potential of the water at any point in the system and μ_0 is the chemical potential of pure free water. The water potential consists of three independent components: a pressure component, a matrix component and an osmotic component:

$$\psi = \psi_p + \psi_m + \psi_0$$

The first component called the pressure potential, represents the actual pressure on the cellular walls and is usually positive in plants even though it may be negative in the vessels. The matrix potential represents that part of specific free energy related to the water in the colloidal structures of the cells and the capillary structure of the soils and is always negative.

Osmosis can be considered as a thermodynamic process. The solvent which has a higher vapour tension than the solution tends to move toward the solution through osmosis or isothermal distillation or diffusion. The water potential gradient is also influenced by the temperature gradient in the plant.

The chemical potential can also be expressed by the equation

$$\psi = RT \ln\left(\frac{p}{p_0}\right), \qquad \text{J mole}^{-1}$$

where R is the gas constant (J mole^{-1} K^{-1}), T is the absolute temperature (K), p is the water vapour pressure and p_0 is the maximum water vapour pressure at a given temperature. This means that the water potential decreases with a decrease in the relative vapour pressure, the latter corresponding to the relative humidity.

Obviously energy coming from the sun is the driving force of the system. Energy balance can be simply symbolised as shortwave radiation $\downarrow$ + longwave radiation

$\downarrow$ + shortwave reflection $\uparrow$ + longwave reflection $\uparrow$ + longwave radiation emitted $\uparrow$ + conversion of energy in vaporizing water + exchange of heat with soil exchange of heat with air = 0.

Also it can be written in another form:

$$R_c = E + R_b + R_c r + K + F + G$$

where R_c is the total incoming radiation, E is evapotranspiration conversion of energy, R_b is back radiation from the surface, $R_c r$ is reflection of energy from the surface (r being the albedo), K is energy spent on heating the air, F is energy spent on heating the soil and G is energy spent on the plant's growth.

Currently used formulae for the calculation of potential evaporation are based typically on the air temperature, air humidity, wind speed, hours of sunshine or net radiation. However, as can be seen from the previous conclusions, there are additional factors relevant to the structure of soils, type of roots and to the mechanism within the plant system which have an equal impact on the formation of evapotranspiration. Therefore all methods of the calculation of actual evapotranspiration based solely on potential evaporation have to be carefully applied.

Another problem recently observed is the seasonal variability of the relationship between potential evaporation and actual evapotranspiration. Considering additional properties of tropical plants such as the ability to shed leaves in critical meteorological situations in order to survive, ratios of actual evapotranspiration to potential evaporation have to be applied very carefully from region to region.

Penman's method of calculating potential evaporation, described elsewhere, is widely used in tropical regions. In addition direct observations provide an additional source of information. Use of the standard Weather Bureau Class A pan for such a purpose under tropical conditions indicates that the metallic pan becomes much hotter than the water it contains and thus measurements are less accurate.

Fig. 3.6 shows a comparison of potential evaporation based on Class A pan observations for the northern part of the tropics and of effective evaporation based on water budgeting, as found by Gonzales and Gauga [23]. In both cases the maximum has been found to be at about 20° south and north of the equator under the most pronounced influence of winds and radiation. In the absence of the meteorological observations from other parts of the tropics, which would be needed for more accurate formulae, the authors tried to develop a realistic formula based on mean daily temperature over 10 °C and regional coefficients which are valid for Cuba:

$$E = 0.53t - 2183, \text{ for } t \geqq 10\ ^\circ\text{C}$$

where E is annual evaporation from free water surface, mm and t sum of mean daily temperatures exceeding 10 °C.

Among many rational formulae, Turc's has been used in the tropics:

$$E = \frac{P}{\left\{0.9 + \left(\frac{P}{L}\right)^2\right\}^{1/2}} \quad \text{mm year}^{-1}$$

where P is annual precipitation (mm), $L = 300 + 25T + 0.05T^3$, T is mean temperature (°C).

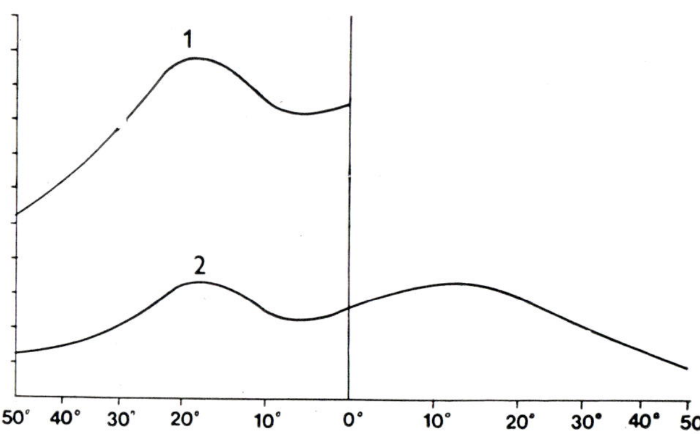

Fig. 3.6 Comparison of evaporation based on Class A pan observation (1) and on budgeting (2), after Gonzales and Gagua

Hydrologists working under tropical conditions have tried to adapt the standartly used pan into a form more suitable for the tropics. An additional attempt was made to protect the water surface from animals and birds. In Africa a fibreglass pan was developed screened by thin wire netting. However, the values of total annual evaporation when compared with the results obtained from the Class A pan were 15% lower. Also the results obtained from the pan protected by special paint to reduce the heating of the pan were significantly different [4].

The process of evapotranspiration from a basin is the result of many effects which are vertically and horizontally variable. For instance in the equatorial rainforest there is a rather high temperature and high humidity at ground level. In the upper layers of the vegetation the influence of the sun and wind is more pronounced and the temperature range is about 8 % and the humidity range 25 %. Ripley [60] presented comparative results of temperature and humidity observation at an elevation of 7.5 metres and 33 metres in the tropical rainforest of Sarawak. The impact of the wind can be judged from the fairly constant humidity in the lower layer compared with the humidity of the upper layer (Fig. 3.7). Similarly the temperature has higher variability. Solar radiation data are not available; however, the total radiation is obviously much lower in the shadowed part of the rainforest.

The profile of the mixed Sasha forest in Nigeria, described by Richards [59] can serve as an indicator of the vertical and horizontal variability of evapotranspiration.

Fig. 3.8 shows five different layers, three of them formed by woods, one by shrub and one by grass. This profile can be compared with the profile of the rainforest described by Beard [7] for Trinidad. In the latter the first layer is very compact and the brush layer is almost absent (Fig. 3.9). Obviously the first type is much

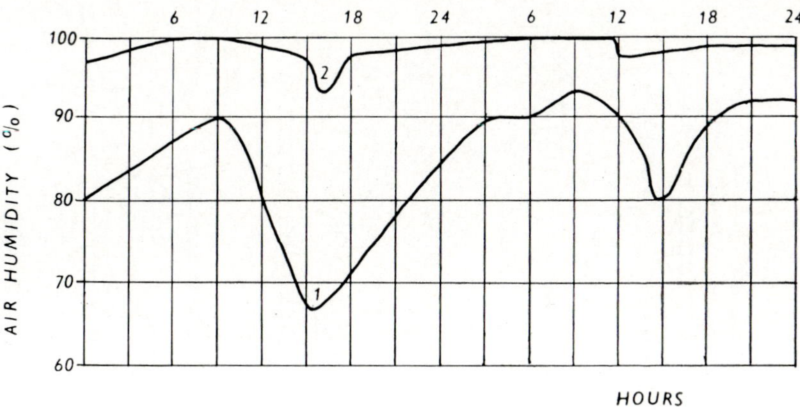

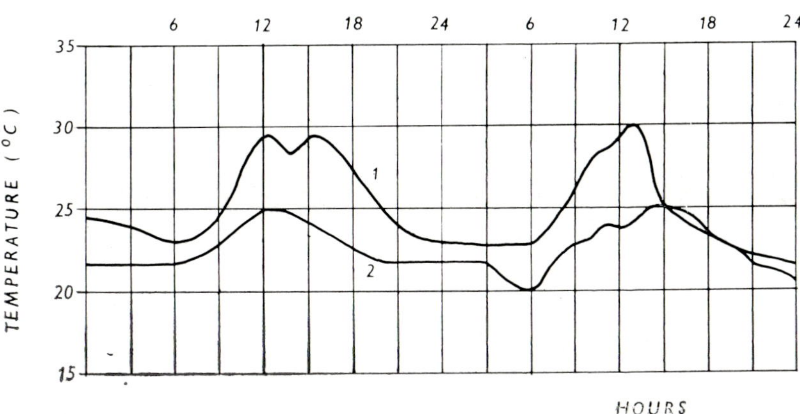

Fig. 3.7 Comparative results of temperature and humidity observations at 7.5 m (2) and 33 m (1) above the ground in Sarawak rainforest

more accessible to radiation and transpiration which at certain times of day can be almost equal at all levels, perhaps slightly reduced by the absence of wind at ground level. In the second example only the upper layer is accessible to full radiation.

Another example is the type of wooded savanna in Ghana as described by Lawson and Jeník [38] in Fig. 3.10. Here the individual woods are accessible to

Fig. 3.8 Profile of mixed Sasha forest, Nigeria, according to Richards.

Fig. 3.9 Trinidad rainforest profile, according to Beard

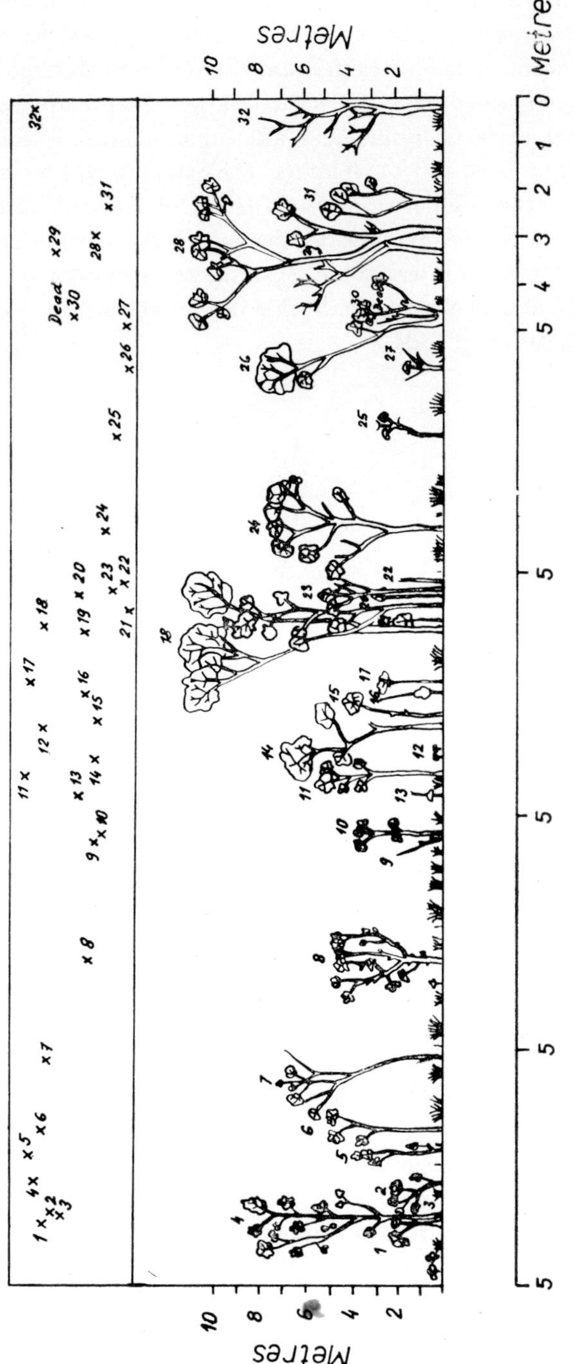

Fig. 3.10 Profile of wooded savanna, Ghana, according to Jeník and Lawson.

evapotranspiration. Actual measurements of evapotranspiration were conducted by the authors on *Securinega virosa,* which is a typical shrub of the Accra Plain and they found significant values of evapotranspiration even during the night hours (Fig. 3.11). From repeated experiments it followed that evapotranspiration was lowest from 18.00 to 06.00 hours and after that increased steadily, achieving a maximum between 12.00 and 14.00 hours. The extent to which the conditions are favourable for transpiration from various levels of wooded savanna where all the layers are more or less exposed to the wind and radiation can be traced on both figures. Nevertheless, it was found that the transpiration rate was lower on the wind-exposed sides of the slump. Such a lack of uniformity was accounted for by the higher dessication of soils.

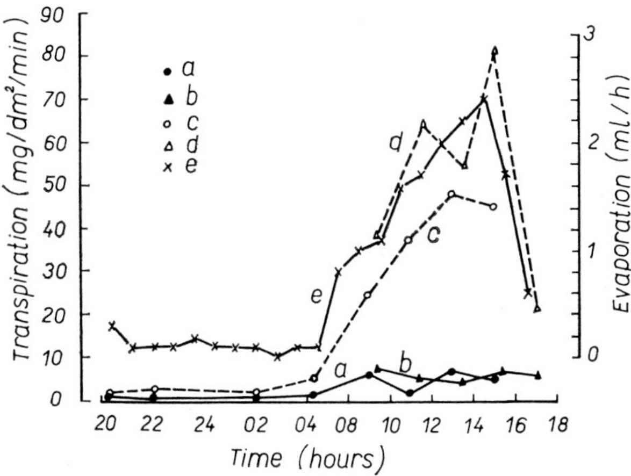

Fig. 3.11 Process of transpiration in Securinega virosa: a, leaves from windward side exposed from windward side; b, leaves from windward side exposed in standard position; c, leaves from leeward side exposed on leeward side; d, leaves from leeward side exposed in standard position; e, upper leaves. Measured 150 cm above ground in grassland, Accra Plains, after Lawson and Jeník

A similar type of experiment was conducted by Balek [5] on the *Brachystegia* species. It was proved that they can easily evaporate 90 % of total precipitation available in a wide range of 600 – 1400 mm per annum. During the rainy season evapotranspiration/potential evaporation ratio E_t/E_0 can easily exceed 1. High variability of total evapotranspiration from watershed to watershed depends on the local composition of the forest and grass because in the wooded savanna evapotranspiration from the area of woods can be three times higher than from the grass.

Hydrologists have long been concerned with the relationship between actual evapotranspiration and potential evaporation. Here, however, the terminology applied is far from being uniform. Sometimes evapotranspiration is considered to be total water loss from the watershed or in other words from vegetation, water surface and soil surface. This obviously also includes evaporation from intercepted water. In some studies evapotranspiration is used as a term for water loss from a specified vegetational cover, perhaps, including the canopy and the vegetation under the main crop. Transpiration is usually used as a term for the water loss from vegetation through the water uptake by the root system. It is not the purpose of this work to provide some kind of standardization of terminology, however, one has to be careful when comparing results obtained in various parts of the tropics.

Many studies presume a constant ratio E_t/E_0 year by year and for the same months of a year. Considering the frequent fluctuation of the groundwater level, the variability of the soil moisture storage in the zone of aeration and large deviations in the precipitation pattern year by year, it is clear that the ratio varies not only day by day but month by month and year by year. Referring to Tab. 3.3 a significant part of E_t is that part of the total evapotranspiration which has been evaporated from the intercepted storage.

The evapotranspiration rate in the tropics can considerably exceed potential evaporation and much higher ratios than those indicated by Tab. 3.4 have been observed or measured. Hurst [32] estimated that Sudanese papyrus may evapotranspire twice as much as a free water surface. Van den Wert and Kammerling [70] reported for water hyacinth E_t/E_0 3.0 to 4.0. Both results, however, were obtained for species which can tape an unlimited amount of water throughout the year and it is quite possible that in the future even higher values of the ratio will be found.

Evapotranspiration from commercial crops was intensively studied on experimental plots in Kenya. Pereira [49] found the ratio for vegetables and pine seedlings to be 0.51, for maize and pine 0.72 and for coffee plantations 0.5 – 0.8, while for the original bamboo cover or forest before the establishment of the plantation it was 0.86 – 0.90. The difference was the result of the increased rate of surface runoff after deforestation. Thus far less of the water infiltrating into the soil layers became available for crops.

A monthly development of the ratio E_t/E_0 was studied by Nieuwolt [47] in Tanzania. The results are shown in Tab. 3.4. Remarkably, ratios higher than 1.0 were found for maize and sorghum when compared with the values for *Eucalyptus*.

A high water loss can be achieved by various types of phreatophytes. Hellwig [29] calculated that about 190 – 500 mm per annum is lost by the phreatophytes *Acacia albedo* and *Tamarix austro-africana* from sandy river beds in Namibia. The water loss would otherwise contribute to the baseflow. The physical background of the process was described by Tromble [68], who studied the water uptake of mesquite trees in a semi-desert region, where the runoff from the watershed headwaters is absorbed by ephemeral streambeds and taken from there by riparian vegetation. A pattern of the groundwater level under such conditions is seen in Fig. 3.12. The vegetation develops in April and May and the water-use rate increases. Thus the water table below the trees decreases as the water-use rate increases to a daily maximum. With decreasing water table elevation the difference between the static head and groundwater level is increased and thus the recharge rate increases. The recharge rate again decreases approximately at 06.00 hours when the difference is at a minimum.

Possibly some other types of mechanism may occur in the evapotranspirational process in semi-arid and arid regions. According to the experiments of Hellwig [29] near Windhoek Namibia with evaporation from moist sand, evaporation was highest just before sunrise. This has been attributed to the difference between the temperature of the water table and that of the air, in the absence of the direct effect of radiation.

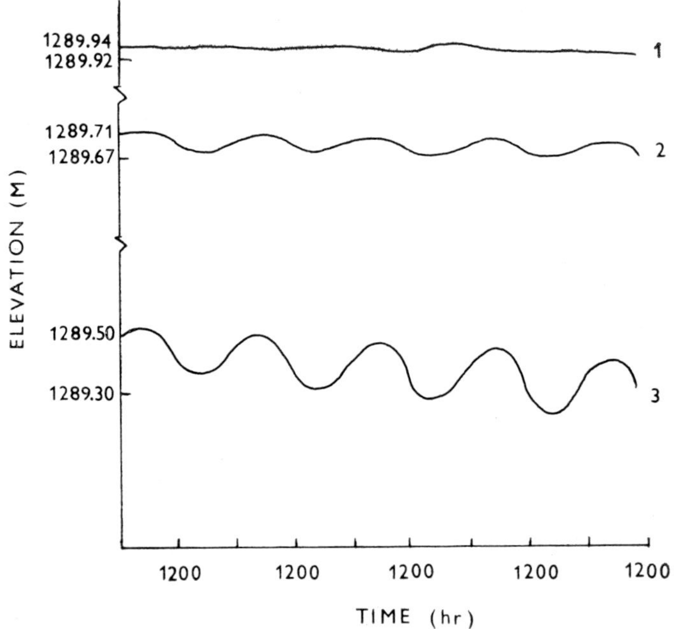

Fig. 3.12 Groundwater consumption by mesquite trees dependent on the season, after Tromble. 1, March; 2, April; 3, May

Grassi [25] in Venezuela tried to develop an empirical relationship between E_t and E_0. The formula resulting from his experiments has the following form:

$$E_t = 0.95E_0C_zC_vF \qquad \text{mm day}^{-1}$$

Here E_t is the daily amount of transpiration (mm), E_0 is potential evaporation, C_z is temperature coefficient and C_v is the vegetational coefficient so that

$$C_z = 1.40 - 0.020T$$
$$C_v = 0.0942 + 0.02774V_c - 0.0002126V_c^2$$

where T is the mean daily temperature, V_c is the percentage of the duration of the vegetational cycle and F is the crop factor (for alfalfa 1.1, for cotton 0.98, for oats

0.83, for beans 0.86, for maize 1.05, for potatoes 1.04, for sugar beet 1.16 and for wheat 0.87).

A more general formula for annual evapotranspiration as dependent on potential evaporation was given by Pike for the tropics:

$$E_t = \frac{RE_0}{\sqrt{R^2 + E_0^2}} \quad \text{mm year}^{-1}$$

where E_t is the annual amount of water deficit (mm year^{-1}), R is the annual rainfall and E_0 is the open pan evaporation or Penman's estimate of potential evaporation (mm year^{-1}).

A comparison of the evaporation from the free water surface of a Class A pan with actual evapotranspiration is plotted in Fig. 3.13. It can be seen that there is a change of pattern in May shortly before the start of the rainy season. A similar change in the evapotranspiration pattern before the rainy season starts was also observed in wet and dry African tropics.

How the generalisation of results achieved for one type of species can be misleading, was proved by Greenwood and Beresford [26], who studied several types of *Juvenile Eucalyptus* in Australia and found a high variability of evapotranspiration in several measured species. The experiment was conducted under similar conditions for all trees and evapotranspiration was measured by a ventilated chamber technique. No significant regression was found between the transpiration per tree and per unit leaf area, nor was leaf dimension found to be a significant factor in the whole process.

Several standard methods for evaporational and evapotranspirational measurements have been applied in the tropics. The most frequent have been widely described by Pereira [50]. A most difficult type of measurement is the separation of transpiration through the plants from other components of transpiration and evaporation. A method which has been applied in simulated semi-arid conditions was developed by Čermák et al. The measuring device consists of a power input generator, a series of thermocouples, electrodes and a recording system. Power is supplied to the tree trunk through five stainless steel electrodes installed in the hydroactive xylem. The temperature difference between heated and non-heated parts of the measured trunk element is measured by eight thermocouples. Two of them are installed at the upper corners of the central electrodes, two at their lower corners and four on either side of the measuring device (Fig. 3.14). The installation is thermally insulated and shaded against direct radiation from the sun.

The equation for the direct calculation of the transpirational flux through a single tree has a form:

$$Q_{wt}^{rec} = \frac{P}{\Delta T}\left(\frac{\Delta t}{d(n-1)\,c_w}\right)O_{bk} \quad \text{kg h}^{-1}$$

where P is the power input (W), ΔT is the temperature difference between heated and non-heated parts of the element (°C), Δt is preselected time interval (sec), d is the distance of the electrodes (cm), n is the number of the electrodes, O_{bk} is the

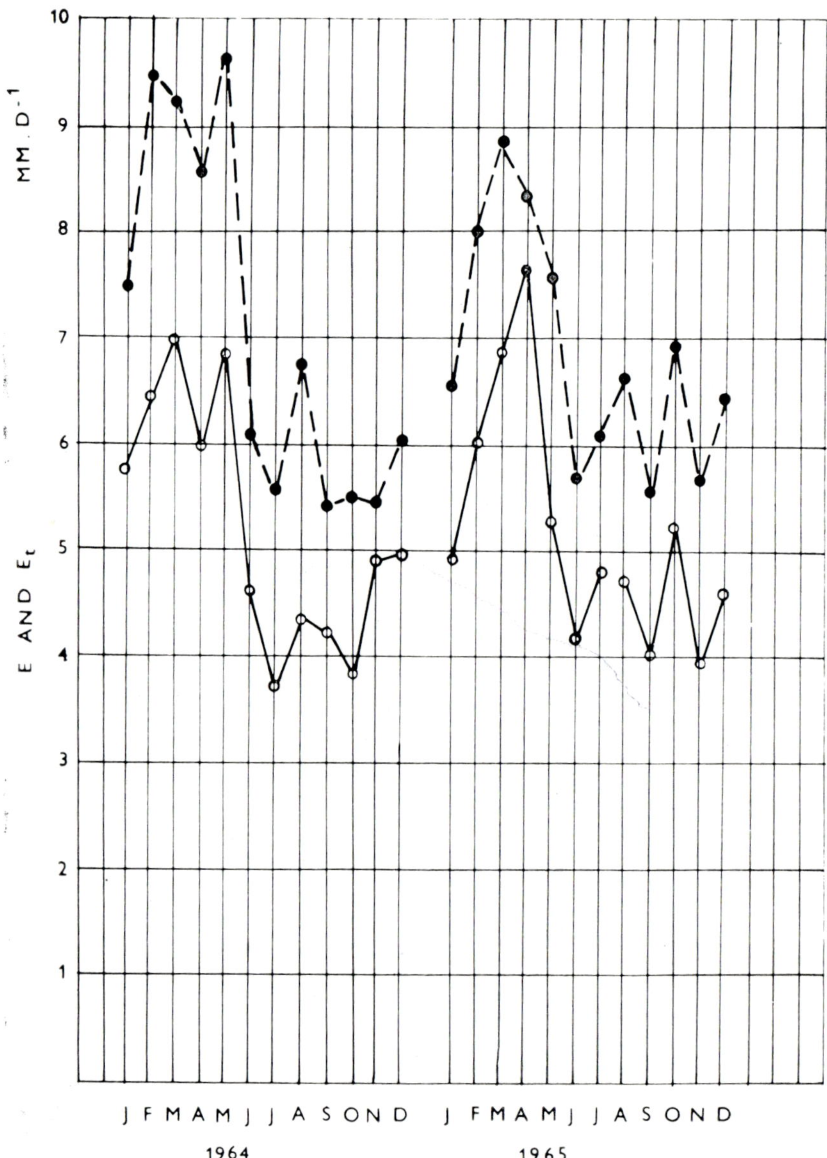

Fig. 3.13 Comparison of evaporation from Class A pan (dotted line), with actual evapotranspiration (full line) in Venezuela, according to Grassi

circumference of the tree without bark at the measured element (cm), and c_w is the specific heat of water, (4186.8 J K^{-1} kg^{-1}).

In the atmosphere-plant-soil water system of tropical plants root systems play a significant role. Under natural conditions roots are well adapted to the prevailing hydrological conditions and also to extreme conditions. Through observation of the depth and density of the roots on natural stands, the hydrologist can gain knowledge about water balance conditions in the region, and about the water exchange between groundwater, soil and plants.

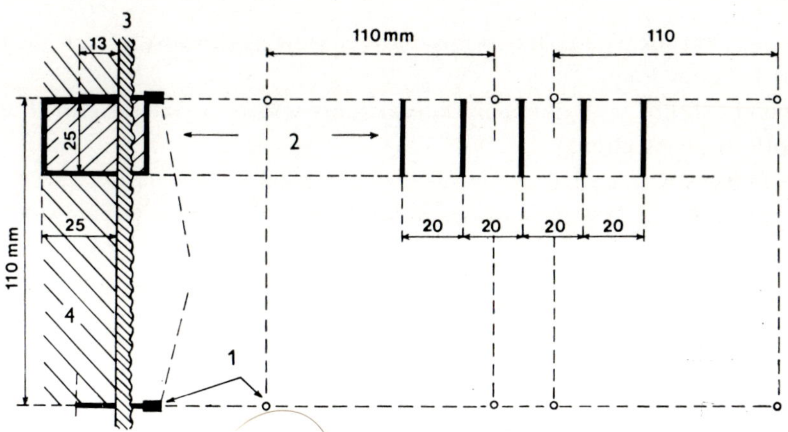

Fig. 3.14 Measuring device for the direct measurement of transpiration according to Čermák, Deml and Penka (Biol. Plant., 15 (3), 1973). 1, thermocouples; 2, steel electrodes; 3, tree bark; 4, active xylem

Biologists differentiate between flat roots penetrating to a depth of up to 1.0 metre, heart roots with a depth of up to 1.5 metres, medium tap roots of up to 2.5 metres and deep tap roots exceeding a depth of 2.5 metres.

For arid zones the following classification of the depth of root penetration has been accepted:

a) annuals with a penetration of no more than 20 cm and laterals within the upper layer of 4 – 5 cm. These roots are typical of many grasses, while for well watered soils the given figures can be exceeded, under natural conditions the roots can be shorter;

b) perennials with their deepest penetration not exceeding 5 metres. They are typical of various cacti. The laterals of these plants extend horizontally to as much as 20 metres. Such a type of root adjustment allows the rapid depletion of soil moisture before it evaporates from the soil;

c) roots extended to the groundwater table level, or at least to the capillary rise zone. Such a system is typical for many plants of wet and dry tropics and for the phreatophytes directly tapping the groundwater. The plants concerned although being a significant source of water loss are of limited economic use. Phillip [51] reported a mesquite taproot with a length of 53 metres. African *Acacias* and *Brachystegias Julbernadias* have similar systems (Maxwell [44]).

Such a classification, based on the observations of Cannon [15] may be found inadequate for many tropical species. For instance, in the case of some tropical grasses a very dense root system has been observed with a total length of eighty thousand metres for a single plant penetrating to a depth of 1.5 metres.

For tropical shrubs three different types of roots have been observed by Hellmers et al. [28]:

a) woody shrubs with coarse primary roots growing downward with the longest taproot 8 metres long,

b) woody shrubs with dominant lateral roots where the lateral growth exceeds the depth of penetration,

c) subshrubs with a shallow fibrous system typical for instance of yucca.

In the first case the system indicates dependence on the deep groundwater table, and in the second on a supply of water from regularly recharged upper soil moisture layers. A rather limited water supply is required by the third type of plant.

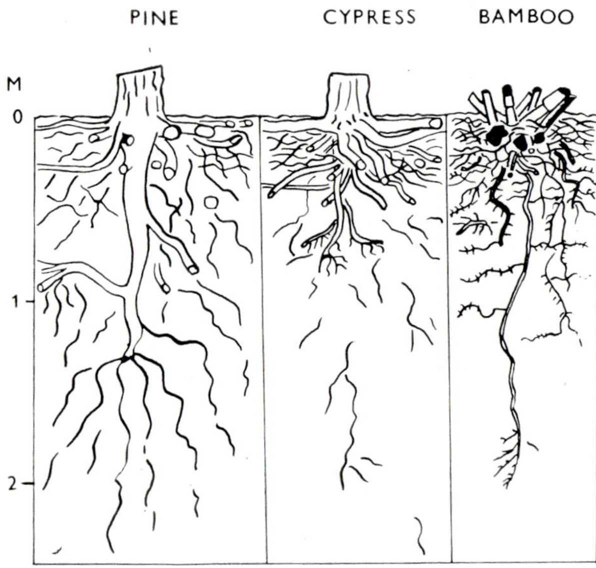

Fig. 3.15 Root system of pine, cypre and bamboo, according to Pereira and Hosegood

Pereira and Hosegood [50] observed root system in experimental catchments in Kenya for the purpose of determining the impact of different types of afforestation on the hydrological regime. They found no significant deviations in the evapotranspiration of pine, cypre or of bamboo, the root pattern of both species being smilar (Fig. 3.15).

The role of the groundwater fluctuation in relation to tree evapotranspiration in the Miombo forest was studied by Maxwell [44]. An analysis of the root system down to a depth of seven metres provided a significant source of information on the function of root density and its relationship to groundwater level fluctuation and soil moisture variability in the unsaturated zone. The trees studied (Fig. 3.16) indicated that the significant part of the root system is located in the upper layers of the soil, tapping easily obtainable water from the zone of aeration, while the secondary system of roots reaches the zone of capillary rise and takes the water through this zone from the groundwater storage during the period of water deficit in upper layers. The experiments were carried out on poorly drained, slowly permeable and strongly acid soils, consisting of sand, loamy sand and sandy loam. The increased density of the roots in the unsaturated zone and in the zone of the capillary rise is clearly visible in Fig. 3.16.

Fig. 3.17 shows the root volumes as distributed under grass. Because the groundwater table is close to the surface and the soil is saturated most of the year, the pattern of this root system is different from that of woody plants.

Fig. 3.18 shows the root systems of various plants in Ghana. The distribution of root systems indicates the soil moisture variability in the region. While the vertical system of the roots of *Cochlospermum Planchonii* indicates that the plants depend mainly on groundwater storage, *Piliostigma Thoningii* takes water from the upper layers of the soil. Other plants have combined root systems similar to those described for the Miombo forest.

Batanoumy and Wahab [6] found that the amount of available moisture in the soil volume exploited by the root system of a small *Laptadenia pyrotechnica* bush amounts to 23 000 kg. The roots penetrate to a depth of 11.5 metres and the whole root system occupies 850 m^3 of soil. The annual water output was estimated to be 5700 kg.

In humid coastal tropics a distinctive ecological region is formed by the roots of mangrove trees. These roots stabilize deposits transported by the rivers to the coast as fast as the material is delivered. The development of roots helps to stabilize the soil and contributes to the formation of new land. With the growth of the roots the land surface rises above what was originally a soil deposit below sea level. Among many types of mangrove three different systems of roots can be recognized (Fig. 3.19). Basically each system has to withstand seawater and fresh water because the swampy regions are flooded twice a day by the high tide. The acceptance of oxygen and release of carbon dioxide results in three different systems of trees:

a) Bruguiera type which form wide shallow rooted loops with branches which are exposed to the air during low tide;

b) Sonneratias also have a wide flat system with vertical off-shots stabilized above the water level;

c) Rhizophoras form still-like roots which support the weight of the tree.

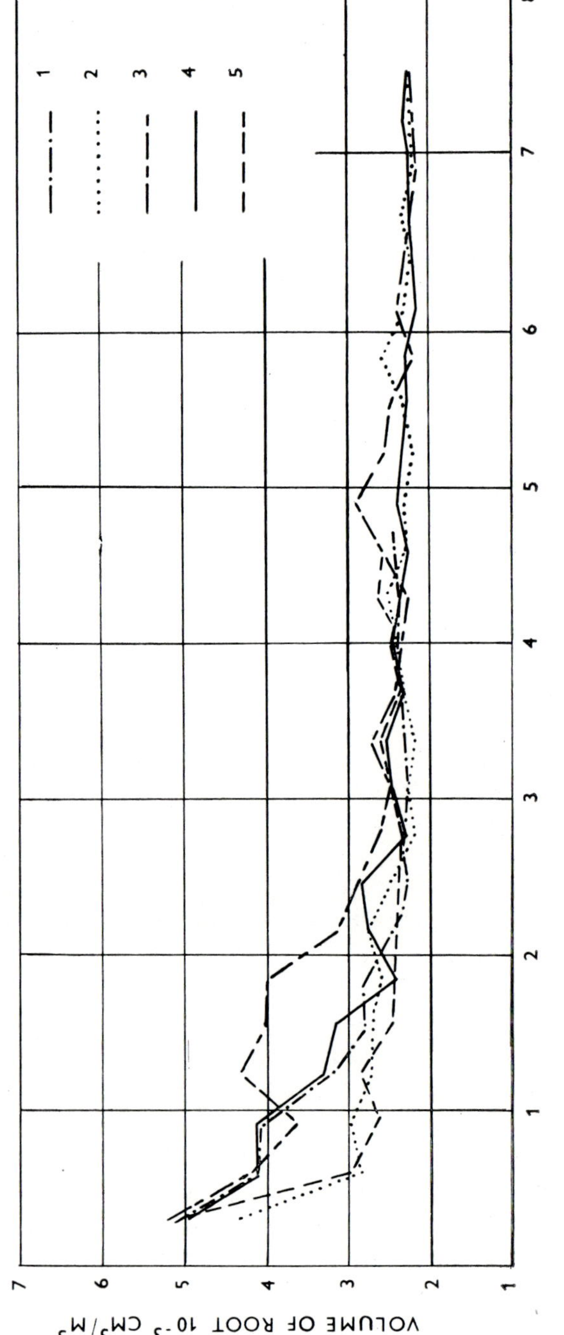

Fig. 3.16 Relation between the root density and soil depth for some species of Central African Plateau, according to Maxwell. 1, *Brachystegia longifolia*; 2, *Brachystegia microphilla*; 3, *Julbernadia paniculata*; 4, *Anisophylles boehmi*; 5, *Baphia bequaerti*

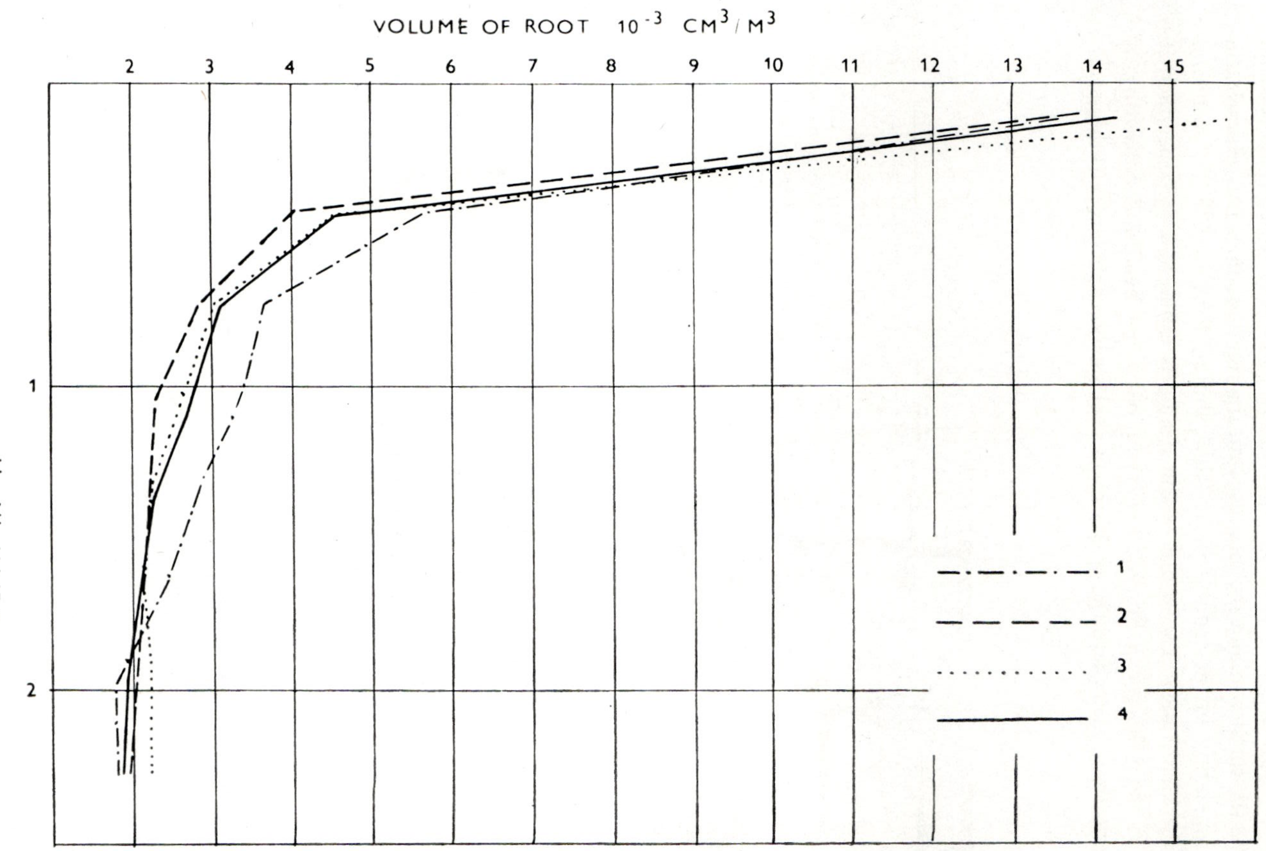

Fig. 3.17 Relation between root density of grass on four catchments of Central African Plateau and soil depth, according to Maxwell

As a general rule the most resistant species form the very front of the coastal region, while the further inland the less resistant the trees become. Thus Sonneratia is found dominantly in the frontal part, Rhizophora forms the central part and Bruguiera the inland part. The width of the system is influenced by other factors and while in the Malayisian Peninsula narrow belts can be found consisting of merely several trees, elsewhere the width of the belts can reach several kilometres.

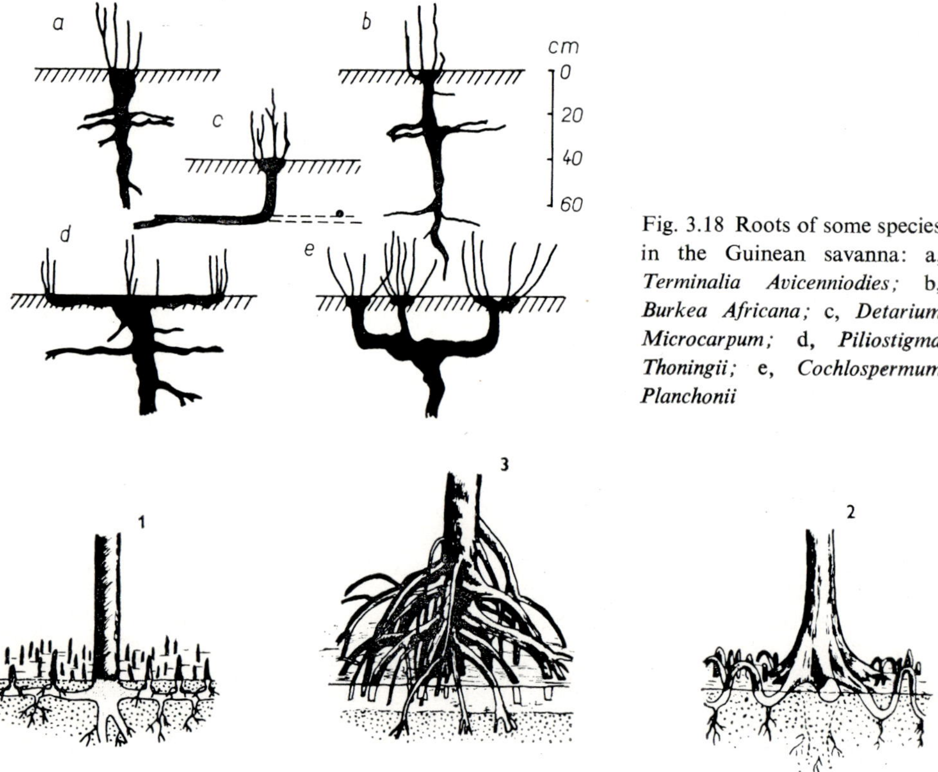

Fig. 3.18 Roots of some species in the Guinean savanna: a, *Terminalia Avicenniodies;* b, *Burkea Africana;* c, *Detarium Microcarpum;* d, *Piliostigma Thoningii;* e, *Cochlospermum Planchonii*

Fig. 3.19 Root systems of mangrove trees: 1, *Sonneratia;* 2, *Bruguiera;* 3, *Rhizophoras*

The process of interception is closely associated with the atmosphere-plant-soil system. The two main roles of interception can be seen in the hydrological cycle. First, the rain falling on the vegetation is transformed into net rainfall which is part of the actual rainfall which creates surface runoff, and second, the water stored in vegetation plays an important role in the process of the evapotranspiration. Water directly evaporates from the leaf surface and because the leaf area is usually more extensive than the area of the gound below the plant, evaporation is extensively enriched through the intercepted water. Water storaged on the surface of leaves also influences the amount of water transpired through a plant, however, very little is known about the actual impact on the whole process.

Another effect which is at present being intensively studied, is the role of interception as a transient factor in the application of fertilisers, pesticides and herbicides. The rate of their application is steadily increasing in many parts of the tropics.

As well as the direct impact of rainfall an equally important role is played by the interception of fog and mist. This process may be regarded as rather insignificant in the tropics, although Ekern [18] calculated that on Norfolk pine in Hawai almost 30 % of the mean annual rainfall was intercepted directly from heavy clouds without any evidence of actual rainfall. Because the mean annual rainfall in the region can reach 2600 mm, a considerable amount of water can be gained by proper selection of the vegetation. Extensive regions of fog forest on the eastern slopes of the Andes gain water support only from the moist passates. The thick layer of fog resulting from warm Pacific winds cooled by the Humboldt stream produces a special type of forest at certain altitude known as loma. The vegetation here is a distinctive product of foggy weather. A similar effect is observed along western parts of the southern African coast.

Intercepted water should not be considered lost for two reasons. First, the rainfall rate of storms is slowed down so that the water can penetrate more easily into the soil and second the cooling effect of the intercepted water reduces the amount of water transpired from the soil.

In contrast, a negative effect of interception is produced when rainfall of small intensity and quantity may not reach the ground at all, nor reduce the transpirational effect.

Pioneering work on interception in the tropical regions was carried out by Jackson [34] and Aston [1]. However, as is often the case their experimental results have limited regional validity and therefore can be used only as a guide when regional problems are to be solved elsewhere. In addition generalization of the effects of interception as existing in various tropical ecosystems is among those fields wide open for research.

Jackson [34] found in Tanzania that the portion of rainfall intercepted increases with the decrease in the actual amount of rainfall. Thus of 1 mm of rainfall 80 % is intercepted, while of 40 mm of rainfall the figure is less than 6 %. Some authors conclude that there is a very slight dissimilarity between the interception of rainfall in jungle, savanna and tropical crops. For the jungle a value between 70 – 80% of the precipitation reaching the soil has been estimated, as being intercepted, for savanna vegetation 80 % and for dense tropical crops 80 – 85 %.

Aston [1] in Australia simulated the rainfall – interception process under artificial conditions. For the experiment he selected eight species common to the Australian tropics: *Eucalyptus viminalis, Eucalyptus maculata, Eucalyptus dives, Eucalyptus cinerea, Eucalyptus pauciflora, Eucalyptus mannifera, Acacia longifolia* and *Pinus radiata*. He found a generalised relationship:

$$I = S\{1 - exp[-(1 - p)P/S]\} + rEt$$

where I is the gross interception loss (mm), p is the proportion of rainfall passing through the canopy (mm), S is the interception storage capacity (mm), r is the ratio of evaporating leaf surface to ground projection area, E is the evaporation rate (mm h^{-1}), t is time duration of storm (hours), P is the gross rainfall (mm).

The proportion of rain passing through the canopy can be simulated by using the leaf area index (LAI) as shown in Fig. 3.20.

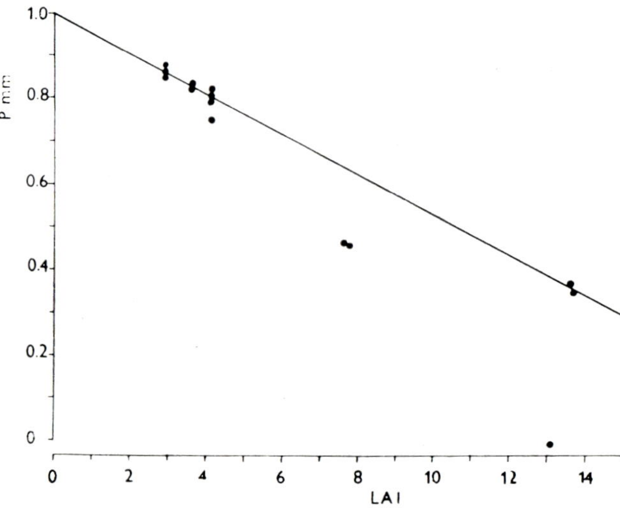

Fig. 3.20 Relationship between the proportion of rain passing through the canopy (P mm) and leaf area index (LAI), for Eucalyptus trees, according to Aston

Balek [4] tested the following equation in Miombo forest:

$$f_i = \left[1 - \left(\frac{\log \frac{w_x}{f_m}}{\log \frac{c_x}{f_m}}\right)\right] f_m$$

where f_i is instant interception rate (mm h^{-1}), f_m is the maximum interception rate which may occur during a year (mm h^{-1}), w_b is the current storage of intercepted water (mm), and c_b is the current capacity of the vegetation to intercept water.

As in the previous formula, the actual values characterizing the vegetational cover have to be found experimentally, or when used in the model, by trial and error. A good estimate for the current interception storage is given by Aston in Fig. 3.21.

Seasonal influence related to the development of leaves and defoliage has been studied by Aston on *Pinus radiata* with remarkable results, rarely available (Fig.

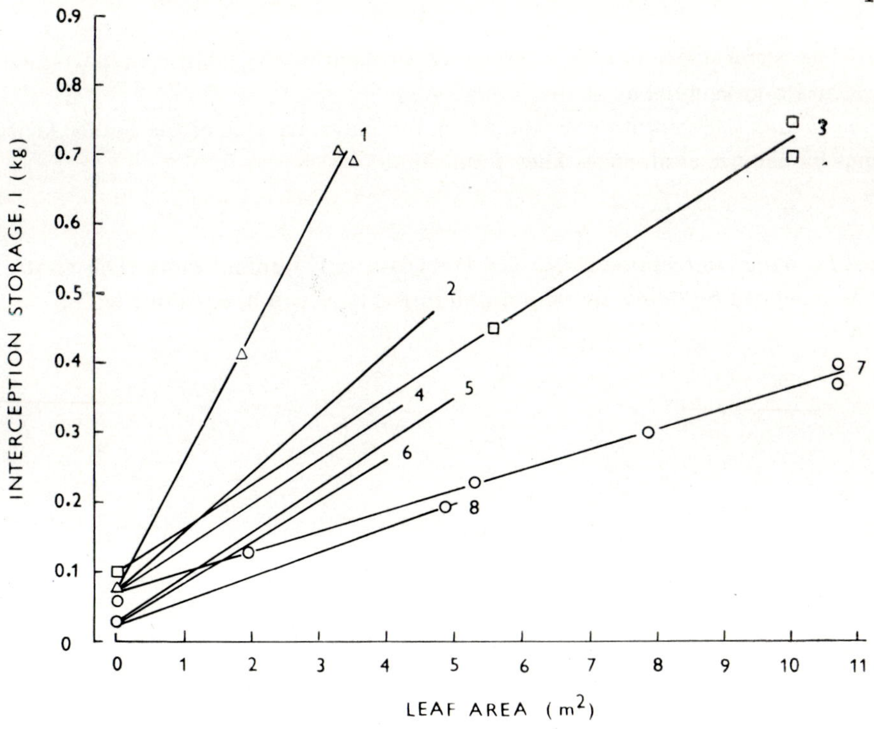

Fig. 3.21 Relationships between interception storage (kg) and leaf area m², according to Aston. 1, *Eucalyptus pauciflora*; 2, *Eucalyptus cinerea*; 3, *Pinus radiata*; 4, *Eucalyptus mannifera*; 5, *Acacaia longifolia*; 6, *Eucalyptus dives*; 7, *Eucalyptus maculata*; 8, *Eucalyptus viminalis*

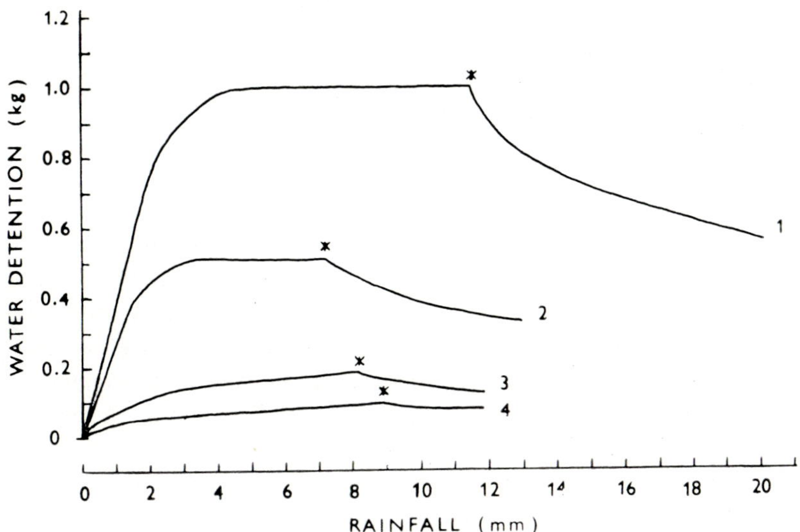

Fig. 3.22 Change in water detention of *Pinus radiata* with rainfall for the whole tree and after three stages of defoliation, according to Aston

3.22). The performance of such a type of experiment is very laborious, particularly the accurate measurement of the foliage area.

Pinus patula interception was studied on the outer margins of the South African tropics by Schulze et al. [62]. They found that

$$I = 0.846 + 2.142 \log P$$

where I is daily interception (mm) and P is gross daily rainfall (mm). The relationship was defined by following the rainfall intensity as can be seen in Fig. 3.23.

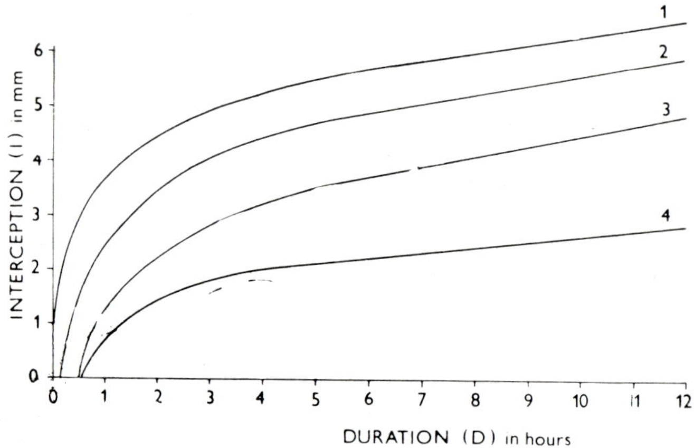

Fig. 3.23 Relationship between rainfall duration and interception for intensity: (1), > 6 mm h^{-1} (2), 3—6 mm h^{-1}; (3), 1.5—3 mm h^{-1}; (4) 0—1.5 mm h^{-1}, after Schulze et al.

3.4 TROPICAL HYDROECOLOGICAL REGIONS

An ecological approach to water management is always emphasized in tropical regions. Fundamental aspects of the structure and functioning and the dynamics of the ecosystems are much more variable here than in temperate regions.

Among many various types of ecosystem classification, hydrologists are anxious for classifications which place emphasis on the water regime and hydrological cycle as affecting the ecosystem on the one hand, and with man's impact on the other. Human activity, however, is highly dependent on the type of ecosystem in which it is practised.

Such an approach can be traced in the works of Gourou [24] and Jarret [35], with further references to a similar type of approach in other works. As can be seen in both works, in an oversimplified classification, we can differentiate between the humid and arid tropics. A solid base for any classification of hydro-ecosystems is the analysis of the impact of seasonal deficits and surpluses of water as related to the biotic and abiotic systems prevailing in the region and also to their gradient conditions. Thus in a more complex classification it can be differentiated between:

a) systems with water deficiency under dry conditions,

b) systems with water excess under wet conditions,

c) systems with combined effects of wet and dry conditions.

This is still a rather simplified scheme not regarding other phenomena, particularly the vegetational cover.

In an endeavour to use vegetation as a dominant ecological factor in South America five major categories were recognized:

a) Desert vegetation, which in Latin America covers land with xerophytic shrub and low drought resistant plants. Scrub forest also invades dry areas in Argentina and certain parts of Mexico. There is also the formation of low dwarf trees called monte. Also considered as belonging to this group are stretches of absolutely barren land, covered with plants after the rare occurrence of rain, heavy dew and fog, as in northern Chile and the coastal region of Peru. Desert vegetation can be generally characterized by the absence of complete cover over the surface.

b) Grasslands and scrub forest are recognized in three types. In cool regions with a slight increase in soil moisture there is short grass steppe, completely covering the ground with sod. This formation is restricted to narrow stretches along the eastern edge of the Andes and to wetter spots in northern Mexico. Where the climate becomes more distinctly humid, tall prairy grass covers the land. This formation covers humid pampa of Argentina, Uruguay and southeastern Brazil. It lacks trees except along the banks of streams. A third type is tropical savanna which covers vast areas of Latin America and especially the Brazilian interior. Wet savanna is found in areas subject to inundation and tall grasses are typical with the absence of trees, while dry savanna is found in well drained areas that are subject to drought. It is mixed with scattered shrub trees, so that it may be difficult to distinguish them from tropical shrub forest. In practically all grassland galerias are to be found that is forest tunnels along the banks of streams.

c) Mediterranean evergreen scrub forest and *brush* with broad leaves is found on the cool side of deserts on the west coast where winter rains and summer droughts prevail. It can be found in northern Chile.

d) Tropical forest is found in many types. Desert borders are frequently marked by tropical scrub forest. Formed by the growth of low scrub trees, broad leaved and deciduous which sometimes grow in thickets with grassy openings, sometimes the grass covers the forest floor. Separation of dry savanna from tropical scrub is not always easy because both areas are composed of scrub trees and tall grasses. The most luxuriant forest, also called selva, prevails in regions where the temperature is never low and rainfall always heavy. Forests with spots of tangled underbrush are regarded as a jungle. Thus both selva and the semi-deciduous forest can be considered as jungle.

e) *Midd-altitude mixed forest* is found in southern Brazil and southern Chile at higher altitudes of mountainous regions. Typically it may consist of a mixture of deciduous species and evergreens.

Tab. 3.5 Classification of the vegetation in the tropical Andes

Months with rainfall	Warm land	Moderate land	Cold land	Frost land
12				
11	Tropical rainforest	Tropical forest	Tropical mountain forest	Paramo
10				
9				
8	Wet savanna	Vegetation of wet valleys	Wet sierra vegetation	Wet puna
7				
6				
5	Dry savanna	Vegetation of dry valleys	Dry sierra vegetation	Dry puna
4				
3	Shrubbed savanna	Shrubbed valley veget.	Shrubbed sierra	Shrubbed puna
2	Semi-desert	Valley semi-des.	Semi-des. sierra	Semi-des. puna
1				
0	Desert	Valley desert	Desert sierra	Desert puna

The classification given in Tab. 3.5 can serve as another approach to ecological classification in South America. Such a type of approach emphasizes one point of view, while the climatological and hydrological aspects become secondary in importance. A modification of this classification (originally made by Kimble [37]) for conditions in Africa can be found to be more hydrologically utilizable:

1. Perennially well-watered regions.

1.1. Rainforest.

1.2. Perennial swamps.

1.3. Great lakes.

2. Seasonally well-watered regions.

2.1. Forest.

2.1.1. Monsoonal tropophilous woodland.

2.1.2. Subtropical type of forest at higher altitudes.
2.1.3. Transitive type of forest.
2.2. Savanna.
2.3. Seasonal swamps and oases.
3. Perennially ill-watered regions.
3.1. Semi-deserts.
3.2. Deserts.

1.1. Any forest in a region where the annual rainfall is in excess of the water needs of plants, even if there is no more water than the forest can use is regarded as a rainforest. Under some conditions the runoff potential can be greater than infiltration into the soil permits. However, in a mature forest there is a continuous downward movement of water and continuous transport of water through the vegetation. After the clearing of the forest either artificially or by fire the soil becomes exposed to solar radiation and erosion and the sensitive balance is easily disturbed. The water table becomes lowered when the changed soil structure reduces infiltration, or raised when the water originally needed for transpiration can infiltrate into deeper layers.

According to Mohr the perfect balance between the production and destruction of humus under luxuriant forest is at a temperature of about 25 °C. It is the forest composition which maintains the soil temperature close to the critical value. Any clearing of the forest results in the destruction of the whole hydro-ecological region.

High temperature and humidity in many parts of the tropics accelerate the spread of a wide range of pests and diseases and encourages the growth of weeds. Often careful treatment of the soil can kill insects and pests through crushing or burying them or bringing them to the surface to be killed by birds or by the sun. The application of pesticides and herbicides on a large scale can be found to be rather expensive in vast tropical rainforest regions.

Rural societies of tropical countries are at a stage of industrial change and a shift from domestic industries to more wide-scale industry can result in additional damage to the forest. While the developing countries benefit from the technological skill and experience of developed nations, this is not the case in pollution control, because this problem has not yet been solved on a world-wide scale.

It would be impossible to list existing species in the rainforest. Among the most significant that can be encountered in Asia are dipterocarpus, palms, climbing ferns, lianas and bamboo, in Africa oil palms, rubber, bamboo, tree ferns and epiphytes, in Australia palms, cypresses, hoop pines, tree ferns and lianas and in South America rubber, palms, castanha, lianas, Andean rainforest with cinchona, tree fern, maquis and montane tropical forest.

1.2. Perennial swamps are found in wet parts of the tropics. In Latin America particularly, swamps tend to be more numerous near the arid tropical margins. From a hydrological point of view they are able to release more water than any other unit, even if not in the most economical way. Much water is lost by evaporation and by the transpiration of aquatic plants. Swamp vegetation plays a very significant role in hydrology and morphology. The effects of swamps are discussed in Chapter 6.

1.3. Lakes occupy much less of the tropics than swamps. The outflow and regime varies from lake to lake and again is dependent on the morphology of each lake system and on climatic factors. Some lakes are without any inflow and/or outflow or they are intermittent and negligible. The hydrologic cycle is complicated in the marginal areas between the lakes and the land where extensive areas of aquatic vegetation transpire considerably. It is not always clear whether these areas belong to the lake or to the land. Sometime the vegetation at the outflow blocks the channel and permanently or temporarily influences the lake regimes. Lakes are also discussed in Chapter 6.

2.1. Forest of seasonally well-watered regions are found in many variations which result from climatic factors and orographic conditions. These regions may experience the additional impact of man's activities.

About half of the tropical regions suffer from alternating surpluses and shortages of water and the ratio of the dry and wet parts of a year is variable. For these regions various types of forest and savanna and mixtures of the two are typical. Walter [73] considered savanna and wet and dry forest as a uniform ecosystem in which grass and woody plants compete for water. In the distinctive forest ecosystems of monsoonal tropophylous woodland in Asia, ironwood, teak, banyan, sandalwood, lianas, bamboo and orchids can be found. In the subtropical forest in the higher latitudes of the tropics oak, camellia, magnolia, rhododendrons, palms, lianas and orchids occur, in Australia woodland with various types of eucalyptus and brigalow scrub with babul, acacia, tamarisk and tamarind, in Africa podocarpus, palms, bananas, lianas, ferns, mosses and epiphytes, in South America there are the so-called 'cerrados'.

The transitive type of forest found at the rims of semi-desert regions consists of xerophilous scrub with acacias and euphorbia in Asia of dry woodland in Africa or sage brush in Central America and Chaco, and xerophilous woodland in Australia.

2.2. Savanna is ecologically defined (Beard [8]) as the natural vegetation on highly mature soils of senile landforms which are subject to unfavourable draining conditions and which have intermittent perched water tables with alternating periods of waterlogging with stagnant water and dessication. In South America a special form of savanna is found on the Andean Plateau, known as a paramo steppe, in Australia savanna with scattered trees and scrub is more typical. For Africa, a savanna with dry woodland or tropical grassland with tall grasses and scattered low trees and bushes, is characteristic, as is grassland with low grass. In South America the grassland is also called llanos and campos.

2.3. Seasonal swamps have various types of vegetation. These include salt swamps which are an indication of intermittent flooding without any significant outflow. Also oases can be ecologically regarded part of this unit. Some of the swamps last for several weeks, other may become temporarily perennial while others are ephemeral. There may also be small pans typical of Australia and dry basins called bolsons in South America from which water disappears with increasing salts at the bottom.

3.1. Semi-deserts experience rainfall only occasionally and usually in a torrential form. The rainwater can rapidly flow underground and become deposited in the river beds or be protected from evaporation if percolation is fast enough. A typical vegetational pattern is found, for instance, in the Kalahari sandveld. Thorn bush, acacia shrub and bunch grasses form the vegetational pattern. In Central America sage brush is found as the dominant cover, while for Asia salt steppes are typical with artemisia, saxaul, acacia, tamarisk. In South America there are semi-desert units of arid tola steppe, Atacama loma, steppes with drought-resisting plants and bunch grasses known as puna and Andean sub-antarctic steppe. In Australia there are dry semi-deserts formed by steppe and malee scrub.

3.2. Deserts have a water supply of, in many cases unknown, exotic origin. Some may travel underground, some may be of ancient origin. In some deserts fog becomes a dominant part of the hydrological cycle, supporting short-lived plants. Similarly night-falling dew during the cooler months is sufficient to support some plants.

Various types of shrubs can be found in some deserts and some of them in the form of salt desert shrubs. In South America Andean ice desert is found as a special feature.

Most tropical ecological types have been subject to man's activity and induced vegetation for a long time and natural systems have been replaced by various types of crops. For instance in a great part of Asia, 95% of the original jungle no longer exists. Some deviations can be also found on the slopes of tropical volcanos where secondary vegetation following a great eruption is not necessarily the same as the original vegetation.

With regard to such features in tropical ecology, Purseglove [56] developed the following classification of the humid tropics:

A. Primary vegetation.

1. Lowland tropical vegetation,

a) well-drained,

b) peat-swamp forest,

c) fresh water swamp forest,

d) riparian forest,

e) mangrove swamp forest,

f) beach forest on sandy and rocky shores,

g) xeroseres,
h) fresh-water aquatic vegetation.
B. Secondary vegetation.
2. Secondary forest,
3. Grassland.
4. Secondary scrub on eroded and exhausted soil.
5. Cultivated land,
a) perennial,
b) annual,
c) swamp cultivation.
6. Urban areas, gardens, parks
7. Devastated areas, mining and building sites

Purseglove strictly objects to the terms jungle and savanna. Other authors have stressed the need to study tropical forest and woodland by means of its groundflora which leads to a more complex classification.

Subrahmanyan [65] studied humid tropics in India and found a difference between per-humid tropics, humid tropics, moist sub-humid tropics and dry sub-humid tropics. In per-humid tropics water deficiency is absent, vegetation has broad leaves and the forest does not always have to be dense and luxuriant. In humid tropics a small quantity of water is left on a seasonal basis for the occurrence of outflow or for a contribution to underground reservoirs. Some deciduousness may indicate seasonal rythms.

In moist sub-humid tropics a recharge of water surplus is combined with a period of deficiency. Typical vegetation is marked as an open jungle with shrubs and tall grasses. A grassland of soft and tall grass types is very rare.

In dry subhumid tropics a continuous swing between deficit and surplus is not typical. More characteristic is the occurrence of resistant types and short coarse grasses. Savanna woodland is most typical, however, it passes into savanna with coarser grasses and slender trees. Dry savanna with thorny trees also exists. A state of deciduousness or dormancy is observed because of prolonged water deficiency. For many species the absence of broad leaves is typical. In general, tropical steppe is less tall and coarse than the wetter savanna of a moist subhumid climate.

A simplified scheme of the world's dominant tropical hydroecological regions can be found in Fig. 3.24.

3.5 WATER BALANCE

A typical water balance problem solved for a tropical watershed can be characterized by a water balance equation

$$P = E_t + R + O + \Delta G + \Delta S + I + M$$

where P is precipitation, E_t is evapotranspiration, R is surface runoff, O is the groundwater outflow from watershed aquifers, ΔG is the groundwater storage increment, ΔS is the soil moisture increment, I is the amount of water intercepted and M is the water recharge or depletion due to the human activity, all values are in millimetres.

An equation of this type can be established for an arbitrary time interval, however, the shorter the interval the more difficult it is to obtain the data, because for shorter intervals higher accuracy of measurement of individual variables is required.

The water balance approach can differ in keeping with the size of the balance area. Basically, we can differ between:

a) a unit of rather small size characterized by a spatially uniform groundwater regime, soil structure, vegetational pattern and orography,

b) hydrological watershed defined by hydrological or hydrogeological boundaries,

c) a subregional or regional unit with boundaries different from hydrological divides, with non-uniform and often complicated hydrological, hydrogeological, ecological, soil and orographic pattern.

Small size units are frequently balanced for various research purposes, as part of agricultural pilot schemes, training programmes, hydrometeorological studies etc. The data observation is concentrated on a small area which can be easily maintained. However, the results are not very representative within a region.

Even the representativeness of a single watershed is often rather limited and some hydrologists counclude that such a type of water balance region does not represent anything except itself. Nevertheless, many usable results have been obtained from experimental and representative watersheds of small size in the tropics, especially when uniformly covered by certain type of vegetation and subject to the changes of the vegetational cover.

The water balance calculations performed for extensive regions, for instance, by the observation of several types of watershed or by an extensive network spread throughout the whole region, is very useful, especially when the data can be collected, stored and utilised through some databank system. However, the maintenance of such a network would be laborious, if performed over a long period.

The precipitation P enters into the calculation as the sum total of rainfall for a specified water balance period. Often a year is used as the water balance period although longer periods of several years or even a decade can be balanced as well. Also shorter balance intervals of one month, day or hour can be used for some special purpose. Precipitation values are relatively easily obtainable, at least at longer than daily intervals. One has to bear in mind that the daily totals are observed at daily intervals for which the time limits skip from one day to the next. It is advisable to work with the water balance values in millimetres, so that the size of the balanced area is eliminated and the results are more easily comparable from area to area.

Evapotranspiration E_t, or as it is incorrectly called 'water loss' in water balance calculation, is among the factors in the equation which are most difficult to obtain. In many cases E_t is the value to be calculated by using the water balance equation, especially when other formulae or direct measurements are not applicable.

Total groundwater outflow O is best obtainable from an analysis of the hydrograph for the balance period through the separation of the baseflow. The methods of separation are described in specialized books. The recession coefficient K is a very useful indicator between the groundwater regime, hypodermic regime and surface regime.

For the decreasing part of the hydrograph component it can be written

$$Q_0 = Q_i K^t$$

where Q_0 is the baseflow discharge after t hours from the initial baseflow discharge Q_i, both values in $m^3\ s^{-1}$ (Fig. 3.25). Hydrograph components can be estimated

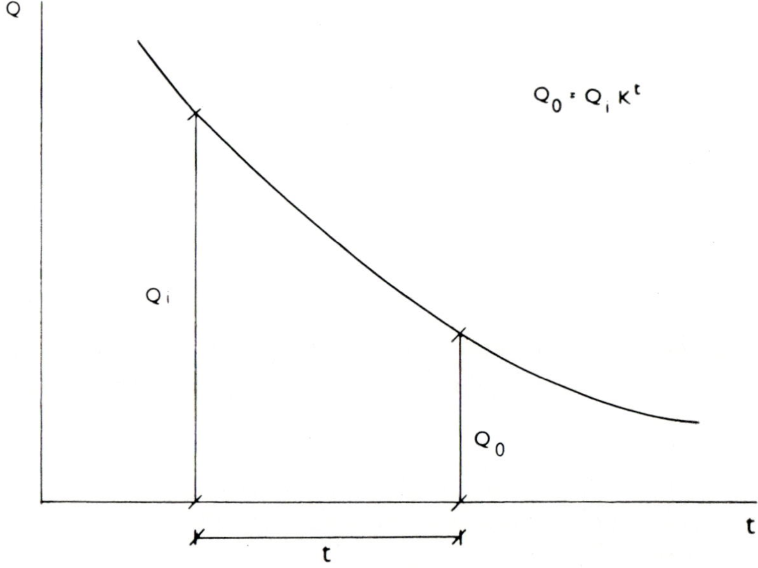

Fig. 3.25 Recession curve

from changing slopes of the decreasing parts of the hydrograph, plotted in semi-logarithmic scale.

Then O is equal to the mean annual baseflow. Alternatively it is equal to the sum of annual baseflows from each contributing aquifer and to the baseflow of hypodermic origin.

Initial groundwater storage involved in the interaction with the baseflow can be calculated by using the formula

$$G_i = \frac{Q_i\,3.6}{A \ln K} \qquad \text{mm}$$

where Q_i is the initial baseflow ($m^3\,s^{-1}$), A is the drainage area from which the baseflow occurs (km^2), and K is the recession coefficient from the previous formula. Then groundwater storage increment ΔG is the amount of water indicating the groundwater storage depletion or recharge when compared with the initial groundwater storage G_i. The fluctuation of water level in the control borehole or well or the final baseflow are the best indicators of groundwater storage development. The mean value from several boreholes provides even better information. Therefore it is essential that the groundwater regime of the selected boreholes should be uniform. A different annual regime of any of the boreholes indicates that it belongs to another aquifer. It can be written:

$$H = \frac{S_g - G}{N} \qquad \text{mm}$$

where H is the mean depth of the groundwater level below the soil surface (mm), S_g is the storage capacity of the aquifer, mm, G is the instant groundwater storage and N is the mean porosity of the non-capillary pores (Fig. 3.26). When calculating

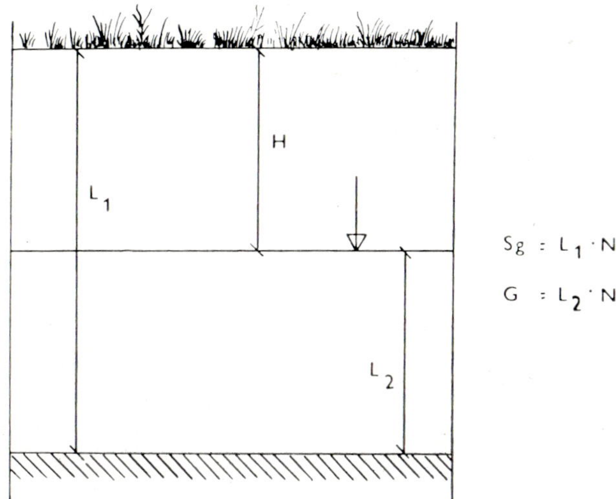

Fig. 3.26 Groundwater storage capacity

G at a selected time interval, the observed H is available and through the comparison of G and G_i, ΔG can be calculated as a positive or negative value.

From the formula it is obvious that uniformity of the soil type is essential when the region is divided into several water balance subregions.

Similarly ΔS or soil moisture increment indicates the change in the water storage in capillary pores as compared with the initial soil moisture storage. Here a direct soil moisture measurement by volumetric methods, by measuring soil resistance or by using a neutron probe is the best indicator of the changes.

I indicates the amount of water which is stored in the vegetation as an interception at the end of the water balance interval. This value is highly variable and difficult to estimate, so the best way is to have an interception equal to zero at the beginning and at the end of the water balance interval.

Last but not least R or the amount of the surface runoff is estimated for a given water balance interval from the hydrograph after the contribution of the baseflows has been deducted. One has to bear in mind that in many hydrographs there is no single baseflow and that the total hydrograph is formed by several types of baseflow originating from different aquifers or from the same aquifer, although under the influence of various different types of vegetational cover. An example of such a composition is given in Fig. 3.27, where the total hydrograph in tropical wet and

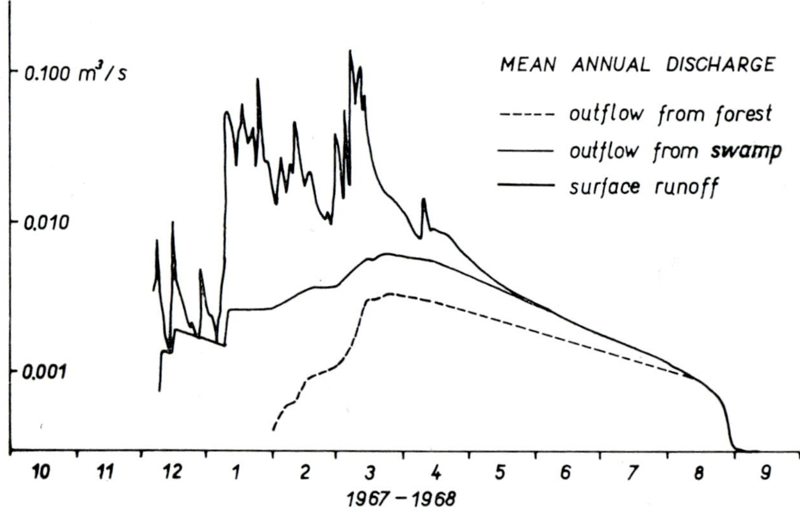

Fig. 3.27 Hydrograph components of the baseflow from the afforested area, from swamp and surface unorff

dry regions is formed by the surface runoff and groundwater flow from the forested area and from the swamp.

Finally M indicates that amount of water which has been exploited from the aquifer or by which the groundwater storage was artificially enriched. A local survey is the main source of information on this variable. In many natural watersheds it is neglected because the water extracted is sooner or later released back into the watershed.

Whether the water balance equation will be solved in simplified or more complex form, depends on several aspects:

a) data availability,

b) length of the water balance interval,

c) ability to identify the region and to divide it into several uniform subregions.

Among many water balance calculations accomplished in various parts of the tropics, the water balance calculation of an exceptionally wet catchment in West Africa, provided by Ledger ([17] Chap. 7) can serve as an example. The balance was calculated for the Guma watershed in Sierra Leone where the values given

Tab. 3.6 Water balance of a wet watershed in Sierra Leone

	Annual rainfall (mm)	Runoff (mm)	Evapotranspiration (mm)	Surplus (mm)
Wet season April-November	5737	4613	572	552
Dry season December-March	58	36	439	—417
Annual	5795	4649	1011	135

in Tab. 3.6 were found in a simplified form. The watershed was considered to be the wettest in Africa and very likely the wettest ever systematically observed in the tropics.

3.6 MATHEMATICAL MODELLING OF THE TROPICAL HYDROECOLOGICAL REGIME

Hydrologists concerned with setting up various types of models simulating parts of the hydrological cycle or their mutual effects in the tropics, do not need to start from the very beginning. Many models have been developed which can be applied directly or with some modifications to tropical regions. Perhaps the latter is more typical, because as has been already stated, many formulae need to be revised before being applied in tropical regions. Some formulae have been generalized only after their application to various tropical problems and testing them under such extreme conditions has often improved the quality of the models. This was the case of the Dambo water balance model (Balek [4]) which was developed from the original 'Guelph model' in such a way that it was able to simulate the behaviour of intermittent headwater swamps, locally called dambos.

Today work with models should not be considered as the main field of tropical hydrology, but rather as a tool or one of the tools for the solution of specific problems. Judgement as to whether the application of a model in a given situation is the best solution, should result from a system analysis approach, which can help to decide on and select the best course of action from several feasible alternatives. An analysis should result from obtaining answers on the following points:

a) required output,
b) identification of available data,
c) identification of the system,
d) data and system analysis,
e) identification of constraints,
f) model selection or rejection,
g) model testing and adaptation.

The required output should result from discussion between the decision maker and the organisation which requests and sponsors the project. Frequently, a request is made to set up a model under any circumstances, even if the situation is not ripe for it or there is no prospect for its further use for any practical purpose. It should therefore be made clear whether the model is to be set up for some more or less scientific purpose in order to confirm or evaluate some hypothetical conclusions or whether the model will serve in its further stage as a tool for managerial decision making, for forecasting etc.

Identification of the data available will have a significant effect on the model work. In the tropics especially abundant data of excellent quality may be found to be available for the pilot scheme, experimental area or in the vicinity of some research centre. This may result in the application of some advanced theoretical model tackling various problems in great detail. However, as soon as the model is extrapolated into a large region, the application of the model can fail because the data are of inferior quality and less numerous.

Also the future development of the network should be considered, together with the future of the existing network used for the development, of the model.

Identification of the system typically consists of the collection of all available maps of orography, settlement, geology, hydrogeology, soil types, vegetation etc. Also all earlier reports concerned with various aspects of the hydrological cycle, ecology, agronomy, engineering development etc., should be reviewed. Information and material collected will indicate ability to describe the behaviour of the system by available formulae, graphs, tables etc., or a need to establish some additional experiments will be signalized.

After the collection of all data and parameters it is possible to provide a basic analysis in order to formulate all interactions involved.

Among typical constraints on model development are the time factor, computer facilities available, limited manpower and budget. Thus a simple model might be found to be the best if only an approximate solution. Much of the time is spent on the basic preparation and preprocessing of the data which can be accomplished by less qualified personnel. However, the persons involved must have a high sense of accuracy and responsibility for the quality of data. The punching of data in a form convenient for the given purpose and suitable for the given computer system is time-consuming work and thus the availability of well-trained personnel must be considered.

Only then can a suitable type of model be selected, or alternatively the model work can be rejected and replaced by perhaps less effective but more realistic standard routines. From the hundreds of models available it may be found difficult to select a suitable one, The simple models with an indication of previous successful applications should be preferred. It is necessary to check whether the model supplies results in a form accessible to persons who may not be quite familiar with the model structure. Also the problems of the compatibility of the model selected to another computer system, which may be available in the future, should be considered.

Among the many models which have been already used and tested, only a small proportion have been used in the tropics. Pitman [52] tested four different rainfall-runoff models of variable complexity on four catchments in the Juksei River basin at Johannesburg, where perfect rainfall – runoff data were available. The author concluded that simple models based on daily and monthly input were quite as satisfactory as more complex models which rely on hourly input. In many parts of the tropics the storage-yield calculations would have to be based on monthly data simply because autographic rain gauges are not available for more complex studies.

An excellent example of how advanced model techniques can be applied in the tropics was given by Bowel et al. [11] in the statistical simulation of the lag response between overland flow, stream discharge and rainfall in a tropical rainforest of northeastern Australia.

One of the largest models applied to the tropical hydrological system was successfully used for the hydrological balance of the upper Nile basin by Brown et al. [12]. Preliminary data collected by a W. M. O. team provided a much needed source of information. The model was set up in order to evaluate alternative plans for water resources development and regulation of the regime in the basins of great African lakes. The model, a description of which can be found in the literature, consists of three submodels, namely a catchment model, a lake model and a channel model. Another positive factor comes from the fact that the model consists of 36 subprograms controlled by a master program and can be used by locally trained staff. However, even such a model cannot be used directly to solve similar problems elsewhere in the tropics. At present, another system of a similar type is being developed for the Amazon basin.

Another hydrologic/hydraulic model has been established for the Indus basin (Zezulák et al. [76]). However, a number of great tropical rivers still remain untouched by sophisticated models.

3.7 REFERENCES

[1] Aston, A. P., 1979. Rainfall interception by eight small trees. Journal of Hydrology 42, 383—396.

[2] Ayyad, M. A., 1976. System analysis of Mediterranean Desert. In Ecosystems of Northern Egypt. Progress Rep. No. 2, University of Alexandria, 200 p.

[3] Bagnold, R. A., 1954. The physical apects of dry deserts. In Biology of Deserts, Editors Cloudsley, Thompson. Institute of Biology, London, 7—12.

[4] Balek, J., 1977. Hydrology and water resources in tropical Africa. Elsevier, Amsterdam, 208 p,

[5] Balek, J., 1974. Hydrology of a tropical forest. (In Czech). Vodohospodářský časopis XXI No. 1, Bratislava, 8 p.

[6] Batanoumy, K. H., Wahab, A., 1973. Eco-physical studies in desert plants. Oecologie 11: 151—161, Berlin.

[7] Beard, J. S., 1946. The Mora forest of Trinidad. Journal of Ecology.

[8] Beard, J. S., 1946. The classification of tropical American vegetation. Journal of Ecology, 89—100.

[9] Bernard, A. E., 1953. L'evapotranspiration annuelle de la forêt equatoriale congolais et son influence sur la pluisité. Comptes Rendus, IUFRO Congress, Rome, 201—204.

[10] Blanford, H. F., 1857. The rainfall of India. Indian Meteorological Memoirs.

[11] Bowell, M., Gilmour, D. A., Sinclair, D. F., 1979. A statistical method for modelling the fate of rainfall in a tropical rainforest catchment. Journal of Hydrology, 42, 251—267.

[12] Brown, J. A. H., Ribeny, F. M. J., Wolanski, E. J., Codner, G. P., 1981. A mathematical model of the hydrologic regime of the Upper Nile Basin. Journal of Hydrology, 51, 97—107.

[13] Brunig, E. F., 1977. The tropical rain forest- a wasted asset or an essential biospheric resource? Univ. of Hamburg Survey Rep. V6, N4, 187 p.

[14] Camaayo, F. C., 1954. Report on the Amazon region. In Proc. of the Abidjan Symp. on tropical soils and vegetation, 11—24.

[15] Cannon, W. A., 1924. General and physiological features of the more aird portions of southern Africa. Carnegie Inst. Washington. Publ. 131.

[16] Dagg, M., Pratt, M. A. C., 1961. Storm from a forested catchment. Hydrol. Symp. Nairobi, 8 p.

[17] Debano, L. F., Thorud, D., B., Schubert, G. H., Gilbert, H., Cochran, P. H., Hunt, J. D., Brown, P. J., 1977. Arid forest land management-problems and solutions. Proc. of the Soc. of Amer. Foresters Natl. Conf., Albuquerque, 74—100.

[18] Ekern, P. C., 1964. Direct interception of cloud water at Lanaihale, Hawai. Proc. of the Soil Sc. Soc. of America 28, 419—421.

[19] Ewel, J. Conde, L., 1978. Environmental implication of any-species utilisation in the moist tropics. USFS/AID Improved utilization of tropical forests Conf., Madison, 63—81.

[20] Furon, R., 1963. The problems of water. Faber and Faber, London.

[21] Gamier, B. J. 1963. Some comments on defining the humid tropics. Research Notes, Ibadan, Vol. 11, 9—25.

[22] Gindel, I., 1964. Transpiration of the Aleppo pine as a function of the environment. Ecology 45, 868—873.

[23] Gonzales, E., Gagra, G., 1979. Nuevo estudio sobre la evaporacion en Cuba. Voluntad hidraulica, No. 51/1979, 5 p.

[24] Gourou, P., 1968. The tropical world. London.

[25] Grassi, C. J., 1968. Estimation de los usos consuntivos de agua y requirimientos de riego. Centro Interamericano de desarollo integral de aguas i tierres, Merida.

[26] Greenwood, E. A. N., Beresford, J. D., 1981. Evaporation from vegetation in landscapes developing secondary salinity. Journal of Hydrology 50, 1981, 155—156.

[27] Gwyne, P., 1976. Doomed jungles? Intl. Wildlife, Jul.—Aug., Vol. 6, No. 4, 36—48.

[28] Hellmers, H., Horton, J. S., Juhren, G., O'Keefe, J., 1955. Root system of some chaparral plants. Ecology 36, 667—678.

[29] Hellwig, J., 1973. Evaporation of water from sand; diurnial variation. Journal of Hydrology 18, 109—118.

[30] Hollis, T., Wood, B., Warren, A., 1979. Rare wetlands in a National Park. Geographical Magazine, London, 331—336.

[31] Holmes, J. W., Williams, A. F., Hall, J. W., Henschle, C. J., 1981. Measurements of discharges from some of the mound springs in the desert of northern South Australia. Journal of Hydrology 49, 329—339.

[32] Hurst, H. E., 1952. The Nile. Constable, London.

[33] Item, H., 1974. Ein Modell für den Wasserhaushalt eines Laubwaldes. Mitt. Eid. Anstalt für das Forst. Versuchwesen. Bd 50, Heft 3, 137—331.

[34] Jackson, I. J., 1975. Relations between rainfall parameters and interception by tropical forest. Journal of Hydrology 24, 215—238.

[35] Jarett, H. R., 1977. Tropical geography. McDonald and Evans, Plymouth, 222 p.

[36] Kilpatrick, P. A., 1980. Simulation of the effects of transpiration losses from the unsaturated zone. Journal of Hydrology 46, 301—309.

[37] Kimble, G. T., 1960. Tropical Africa. 20th Century Foundation, New York.

[38] Lawson, G. W., Jeník, J., 1967. Microclimate and vegetation in the Accra Plains. J. Ecol. 55, 777—785.

[39] Lawson, G. W., et al., 1968. A study of a vegetation in Guinean savanna and Mole game reserve, Ghana. J. Ecol. 56 505—522.

[40] Livingston, B. E., 1907. Relative transpiration in the cacti. Plant World 10, 110—114.

[41] Logan, R. F., 1968. Causes, climates and distribution of deserts. In Desert Biology, Ed. Brown, G., W., Academ. Press, New York, 21—50.

[42] Marinelli, J., 1980. Eco-crime on the equator. Environ. Action, Vol. 11, No. 9, 4—14.

[43] Matějka, V., Pokorný, J., Borota, J., Parkán, J., Martinů, M., 1971. Forest management in the tropics and subtropics. (In Czech). Agricultural College Prague-Textbook, 307 p.

[44] Maxwell, D., 1972. Root range investigation. TR 24 Rep., NCSR Lusaka, 15 p.

[45] Medina, E., Herrera, R., Jordan, C., Linge, H., 1977. The Amazon project of the Venezuelan Inst. for Scientific Research. Nature and Resources-Unesco, Vol. 13, No. 3, 4—7.

[46] Mooney, E., 1976. Inferno verde. Defenders, Vol. 51, No. 1, 24—27.

[47] Nieuwolt, S., 1973. Rainfall and evaporation in Tanzania. B.R.A.L.U.P., Pap No. 24, Dar es Salaam, 48 p.

[48] Nigh, R. B., Nations, J., 1980. Tropical rainforests. Atomic Scientists, Vol. 36, No. 3, 12—20.

[49] Pereira, H. G., 1962. Hydrological effects of changes in land use in some East Afr. catchment areas. E.A. Agr. For. Journal 24.

[50] Pereira, H. G., 1973. Land use and water resources in temperate and tropical climates. Cambridge Univ. Press, 246 p.

[51] Phillip, W. S., 1963. Depth of roots in soil. Ecology 44, p. 424.

[52] Pitman, W. V., 1978. Flow generation of catchment models of differing complexity. Journal of Hydrology 38, 59—70.

[53] Pitman, W. V., 1978. Trends in streamflow due to upstream land-use changes. Journal of Hydrology 39, 227—237.

[54] Poore, D., 1978. Values of tropical moist forest. USFS/AID Improved Util. of Trop. For. Conf., Madison, 39—62.

[55] Potter, G. L., Ellsaesser, H. W., MacCracken, M. C., Luther, F., M., 1975. Possible climatic impact of tropical deforestation. Nature, Vol. 258, No. 5537, 697—699.

[56] Purseglove, J. W., 1956. A note on the vegetation of humid tropics with special reference to Singapore. In "Study of trop. veg.", Kandy, Ceylon, Unesco, 84—88.
[57] Ranjitsinh, M. K., 1979. Forest destruction in Asia and the South Pacific. Ambio, Vol. 8, No. 5, p. 192.
[58] Richards, P. W., 1939. Ecological studies on the rainforest of southern Nigeria. J. Ecol. 1939.
[59] Richards, P. W., 1957. Tropical rain forest. Cambridge Univ. Press.
[60] Ripley, S. D., 1966. The land and wild-life of tropical Asia. Time-Life Amsterdam, 199 p.
[61] Savory, C. R., 1965. Game utilisation in Rhodesia. Zoologia Africana 1, 321—337.
[62] Schulze, R. E., Scott—Shaw, C. R., Nänni, W. W., 1978. Interception by Pinus patula in relation to rainfall parameters. Journal of Hydrology, 36, 393—396.
[63] Smith, H. T. W., 1968. Geologic and geomorphologic aspects of deserts. In "Desert Biology", Ed. Brown, G. W., Academic Press, N. York, 51—100.
[64] Smith, H. T. W., 1977. Status of desertification in the arid regions. UN Conf. on desertification, Nairobi.
[65] Subrahmanyam, V. P., Murty, P. J. N., 1968. Ecoclimatology of the tropics with special reference to India. In "Rec. Adv. in Trop. Ecol.", Ed. Misra., R. Hindu Univ., Varanasi.
[66] Taylor, Ch. J., 1962. Tropical forestry with particular reference to West Africa. London.
[67] Teller, H. L., 1977. Environmental impact analysis and forestry activities. In "Guidelines for Watershed Management", FAO, Rome, 15—25.
[68] Tromble, J. M., 1977. Water requirements for mesquite. Journal of Hydrology 34, 171—179.
[69] Tromble, J. M., 1973. U.N.D.P., Hydrometeorological survey of the catchments of Lakes Victoria, Kyoga and Mobutu, Entebbe.
[70] Van den Wert, P., Kamerling, G. E., 1974. Evapotranspiration of water hyacinth. Journal of Hydrology 22, 201—212.
[71] Veblen, T., 1976. The urgent need for forest conservation in Guatemala Highland. Biol. Conserv., Vol. 9, No. 2. 141—155.
[72] Went, F. H., Babu, V. R., 1978. Plant life and desertification. Env. Conserv., Vol. 5, No. 4, 263—272.
[73] Walter, H., 1964. Productivity of vegetation in arid countries. Ecol. of Man in Trop. Evironment, Mogens, Switzerland, 9 p.
[74] Whyte, R. O., 1966. The use of arid and semiarid lands. Proc. of Unesco Symp. on Arid Lands, 8 p.
[75] Wicht, C. C., 1949. Forestry and water supplies in South Africa. Dep. Agr. S. Afr. Bull, 58, 33 p.
[76] Zezulák, J., Gabriš, P., Drako, J., Martínka, K., Svoboda, A., 1980. Flood warning and forecasting system in Indus river basin. U.N.D.P./W.M.O. Project, Pak. 74/027.

4 RIVERS AND BASINS

4.1 GENERAL

Seven of the ten largest rivers in the world originate in the tropics, namely the Amazon, La Plata, Congo, Madeira, Orinoco, Tocantins and Rio Negro. The powerful streams of the Madeira, Tocantins and Rio Nego are tributaries of the Amazon and therefore the list of the main tropical rivers can be extended as indicated in Tab. 4.1. Here the Mekong, Irrawaddy, Niger, Zambezi, Nile and Sao Francisco join the family of the ten largest tropical rivers. The Irrawaddy has its headwaters outside of the tropical boundaries and the lower reaches of the Nile are outside tropical limits, nevertheless, both rivers are products of tropical conditions.

Tab. 4.1 Great tropical rivers

No.	River	Mean annual discharge ($m^3\ s^{-1}$)	Drainage area (km^3)	Yield ($l\ s^{-1}\ km^{-1}$)	Length (km)
1	Amazon	212 000	6 140 000[1])	34.5	7025
2	La Plata	42 400	3 104 000	13.7	4580
3	Congo	38 800	3 607 450	10.7	3400
4	Orinoco	28 000	1 050 000	28.0	2400
5	Mekong	13 500	810 000	16.6	4200
6	Irrawaddy	13 400	431 000	31.0	2150
7	Niger	7000	1 091 000	6.4	4160
8	Zambezi	3800	2 200 000[2])	1.7	3250
9	Sao Francisco	3800	650 000	5.8	3720
10	Nile	2590	2 881 000	0.9	6500

[1]) 6 915 000 km^2, according to World water balance [40], 7 050 000 km^2, according to Karasik [20].

[2]) Including part of Kalahari.

Because these rivers and their basins are under the influence of very complicated regimes, the term 'typical river regime' cannot be related to any of them. Such a characterization is pertinent to small and medium-sized streams even if many of them are influenced by certain local conditions affecting the river regime to a greater or lesser degree.

The rivers in Tab. 4.1 are listed according to their annual discharge. If river length or the drainage areas were considered as the main features a different order would be obtained. The water balance of the great basins can be seen in Tab. 4.2.

Tab. 4.2 Water balance of great tropical basins

River	Precipitation (mm)	Runoff (mm)	Evapotranspiration (mm)	Runoff coefficient (mm)
Amazon	2150	1088	1062	0.51
La Plata	1240	432	808	0.35
Congo	1561	337	1224	0.22
Orinoco	1990	883	1107	0.44
Mekong	1570	523	1047	0.33
Irrawaddy	1970	978	992	0.50
Niger	1250	202	1048	0.16
Zambezi	750	54	696	0.07
Nile	506	28	478	0.06
Sao Francisco	1050	183	867	0.17

4.2 CLASSIFICATION OF TROPICAL RIVERS

Many atempts have been made to classify river regimes. Voeykov [39] considered rivers to be products of climate and the climate remains a decisive factor in the majority of classifications. Voeykov classified all tropical rivers together as being fed by rainwater and having their highest discharge during the summer season.

Pardé [28] recognized:

a) tropical rivers characterized by a minimum discharge during the winter season and maximum discharge during the summer season,

b) equatorial rivers with two peaks.

Lvovich differentiated in the tropics between:

a) rivers fed only during the autumn such as the Congo and the White Nile,

b) rivers fed mostly during the autumn such as the Niger and the Nile,

c) rivers fed during the summer.

Sanchez [33] in Mexico recognized rivers with pluvial, glacial and mixed regimes, his classification being based on extensive records from the beginning of the twentieth century.

With the increasing density of the observation network new regimes have been found to be typical for parts of the tropics. Perhaps the greatest variety of river regimes are found in the South American tropics, where the impact of zonal stratification in the Andes is one of the most decisive factors.

Netopil [26] recognized basic river regimes in South America as:

a) equatorial rivers of the Amazon basin with rainfall throughout the year. One or two maxima appear according to the sun culmination and steady flow month by month is a dominant feature,

b) tropical rivers with higher discharge fluctuation throughout the year. During the drier period the rivers are fed only by the baseflow. A single flood wave influences the regime. Great tributaries of the Amazon and Orinoco are examples of this type of regime,

c) rivers fed by the combination of seasonal rainfall and the rain bringing effect of trade winds. These rivers are found in tropical parts of southern Brazil and Paraguay,

d) rivers without pronounced dry and wet periodicity with irregular floods rapidly decreasing during the dry season and with a prolonged period of low flow. Such rivers are found in northeastern Brazil and northern Venezuela,

e) rivers of dry regions fed partly by rainfall and melted snow. Discharge amplitude is even more pronounced than in the case of the previous regime and some of these rivers are intermittent or ephemeral,

f) rivers fed by snowmelt and by the glaciers of the tropical Andes. Variability during the year is not too pronounced, daily variability is often more typical. Rivers of this kind are found in the headwaters of the Amazon and Marañon.

g) rivers fed by fog condensation in the dry belt of the western Andes. The flow is low and diminishing in the desert,

h) transitive rivers under various combined influences,

i) rivers fed dominantly by the baseflow from mountainous regions with intensive weathering.

A classification based on the observation of Australian rivers can be found in Soviet sources [40], namely:

a) rivers of the arid zone with eastern, northern and western subtypes,

b) rivers of the humid coastal zones with northwestern, eastern, southeastern and southwestern subtypes,

c) rivers of the semi-arid subtypes with northern, eastern and southern subtypes,

d) rivers of tropical islands.

An earlier type of classification based on hydroecological regionalisation (Balek [47]) has been extended to apply to all parts of the tropical continents:

1 Equatorial rivers of the humid tropics.

1a Equatorial rivers with one peak produced by heavy annual precipitation of over 1750 mm, without a pronounced dry season. Often the rainfall pattern has two periods of increased precipitation, however, one flood peak is much higher and corresponds to the peak of precipitation. Fig. 4.1 and Tab. 4.3.1 give the Brantas River at Pohgadjig, Indonesia as an example.

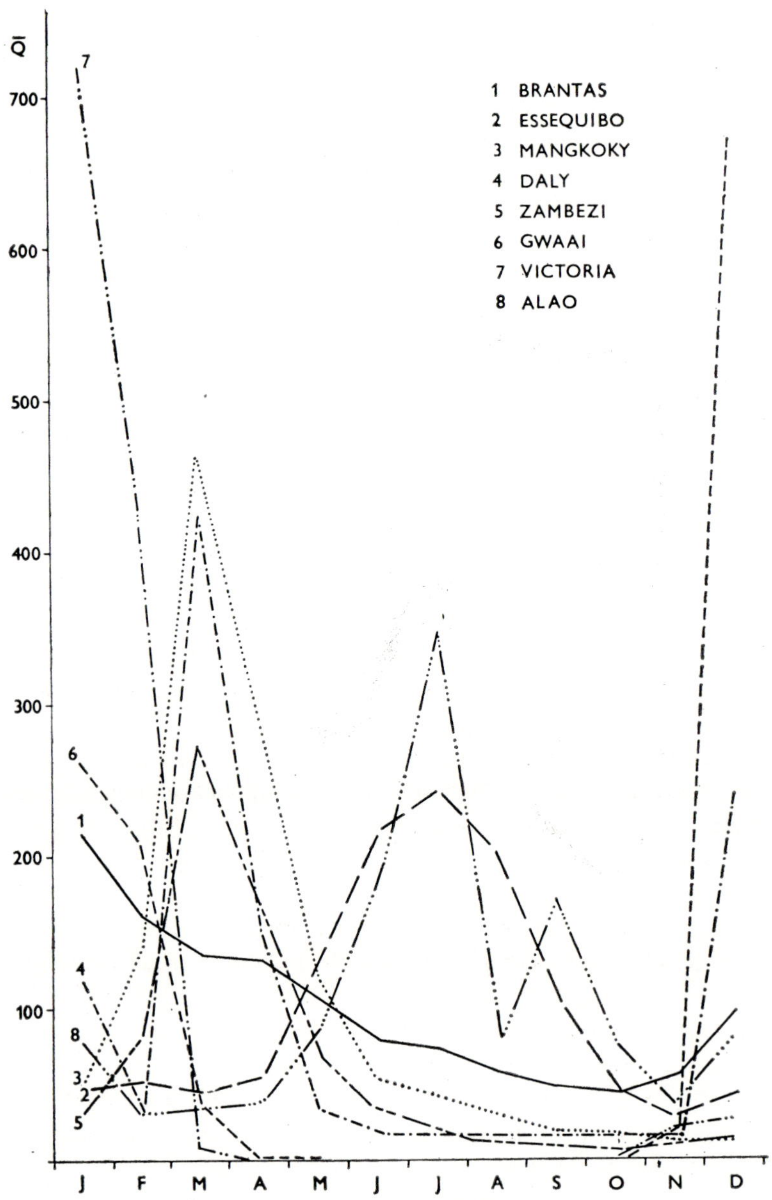

Fig. 4.1 Main types of the river regime in the tropics

Tab. 4.3.1 Brantas basin at Pohgadjih, Indonesia (8°0′ S, 112°0′ E, 2176 km²)

	J	F	M	A	M	J	J	A	S	O	N	D	Annual
Rainfall (mm)	321	286	268	196	107	89	54	18	17	54	125	250	1785
Discharge ($m^3 s^{-1}$)	108.7	82.0	70.0	67.8	55.2	39.4	36.1	29.5	23.9	20.8	27.5	49.3	50.8

1b Equatorial rivers with two peaks produced by precipitation regimes with monthly totals over 100 mm. The basins are dominantly covered by tropical forest with an annual precipitation of over 1750 mm. The River Essequibo in Guyana is given as an example (Fig. 4.1.2 and Tab. 4.3.2).

Tab. 4.3.2 Essequibo basin at Plantain Island, Guyana (5°50′ N, 58°35′ W, 68 600 km²)

	J	F	M	A	M	J	J	A	S	O	N	D	Annual
Rainfall (mm)	266	133	161	165	346	372	346	186	106	100	160	319	2660
Discharge ($m^3 s^{-1}$)	1040	1120	998	1180	2790	4770	5320	4450	2270	979	698	889	2190

2 *Rivers of wet and dry regions.*

2a Rivers of tropical wet and dry lowlands of a relatively moist type. They are found in areas with a moist type of savanna and where the seasonal effect of rainfall is pronounced. The dry season persists for at least three months. The River Mangkoky in the western Malagasy is given as an example (Tab. 4.3.3 and Fig. 4.1.3). Although there are four dry months, there is always some precipitation in

Tab. 4.3.3 Mangkoky basin at Banian, Malagasy (21°50′ S, 44°01′ E, 50 000 km²)

	J	F	M	A	M	J	J	A	S	O	N	D	Annual
Rainfall (mm)	295	240	235	57	6	0	n	n	n	45	191	179	1248
Discharge ($m^3 s^{-1}$)	554	977	3234	1970	834	379	205	128	83	49	64	120	700

the basin and the streams never become intermittent. However, the fall of the hydrograph towards the end of the dry season is rather steep.

2b Rivers of the tropical wet and dry lowlands of a relatively dry type. The basins are covered by the dry type of savanna and the seasonal effect of rainfall is well pronounced. The dry season is prolonged and continues for several months so that the rivers, especially with smaller basins may occasionally behave as intermittent. The rainfall is below 1000 mm. An example of this is the river Daly in northern Australia (Fig. 4.1.4 and Tab. 4.3.4) which in the long term mean always has some monthly discharge, but which occasionally dries out during the dry season.

Tab. 4.3.4 Daly basin at Gourley, Australia (14°0′ S, 130°50′ E, 46 300 km²)

	J	F	M	A	M	J	J	A	S	O	N	D	Annual
Rainfall (mm)	176	135	109	26	6	3	1.5	1.5	13	19	37	97	644
Discharge ($m^3 s^{-1}$)	74	33	264	152	21	12	10	9	8	8	9	148	62

2c Rivers of the tropical wet and dry highlands of a relatively moist type have their basins covered with woodland and moist savanna. The length of the dry season varies from place to place. The headwater may receive more than 1000 mm of the annual rainfall, while the tributaries of the middle and lower reaches receive less than 1000 mm. The upper Zambezi at Chavuma Falls is given as an example in Fig. 4.1.5 and Tab. 5.3.5. The dry season is shorter than in the lowlands, although there are at least three dry months without any rainfall.

Tab. 4.3.5 Zambezi basin at Chavuma Falls, Zambia (13°05′ S, 22°41′ E, 29 331 km²)

	J	F	M	A	M	J	J	A	S	O	N	D	Annual
Rainfall (mm)	295	240	235	57	6	n	n	n	n	45	191	179	1248
Discharge ($m^3 s^{-1}$)	354	977	3234	1970	834	379	205	128	83	49	64	120	1161

2d Rivers of the tropical wet and dry highlands of a relatively dry type. These rivers are found in the marginal parts of the wet and dry climate zones. In general, the basins receive little more than 500–700 mm of rainfall. The basins are dominantly covered by dry savanna. The river Gwaii, Zimbabwe is given as an example in Fig. 4.1.6 and Tab. 4.3.6. Although the drainage area is almost 40 000 km^2 in some years there is no discharge at all.

Tab. 4.3.6 Gwaii basin at Kamativi, Zimbabwe (18°22′ S, 27°03′ E, 38 632 km^2)

	J	F	M	A	M	J	J	A	S	O	N	D	Annual
Rainfall	101	114	76	25	2	1	0	0	0	0	237	145	711
Discharge ($m^3 s^{-1}$)	15	12	2.14	0.19	0.007	0	0	0	0	0	1	38	5.7

3 Dry climate rivers.

3a Semi-desert climate rivers are influenced by an extended dry season. Ecologists differentiate in the dry semi-desert regions between wooded steppe and grass steppe. Different types of vegetation have very little impact on the hydrological regime however, and it is rather the size of the drainage area that becomes more significant. Small streams may show as intermittent behaviour more frequently than larger ones. The annual rainfall is below 500 – 700 mm. Shallash and Starmans developed a relationship between the annual rainfall, runoff and drainage density indicating that the drainage density is a good indicator of the water yield in the basins in semi-deserts:

$$Q = 0.001\,35 \frac{R^{3/2}}{D}$$

where Q is the mean annual streamflow (cfs), R is the total annual amount of rainfall over the catchment (in), and D is drainage density (mile mile^{-2}). The Victoria river, Australia which behaves as a periodically intermittent stream (Fig. 4.1.7, Tab. 4.3.7) is given as an example.

Tab. 4.3.7 Victoria basin at Coolibah, Australia (15°40′ S, 131°0′ E, 44 900 km^2)

	J	F	M	A	M	J	J	A	S	O	N	D	Annual
Rainfall (mm)	137	112	79	14	5	5	5	n	n	14	34	83	488
Discharge ($m^3 s^{-1}$)	529	323	6	1	0	0	0	0	0	0	15	17	74

A similar attempt to classify the water yield according to the density of the drainage network was made by Daniel who found a linear relationship between the drainage density and annual precipitation in Guyana (Fig. 4.2) and concluded that a relationship exists between the drainage density and climatic factors. Over 3000 mm of the annual rainfall the drainage density remains more or less constant and there is also some difference between the drainage density of the rivers in savanna and forest.

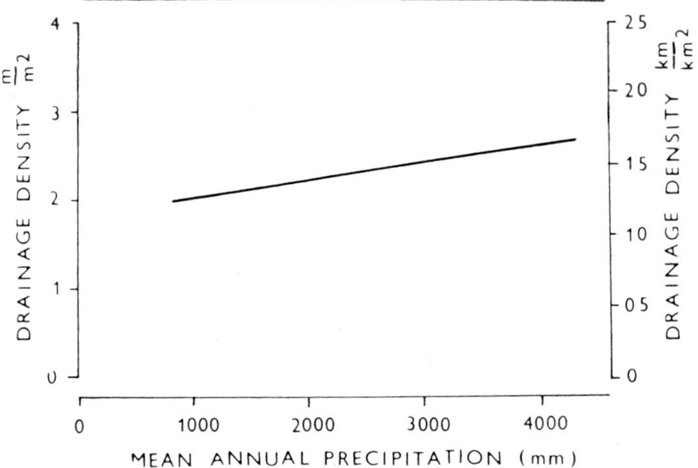

Fig. 4.2 Linear relationship between the drainage density and annual precipitation in Guyana, according to Daniel

3b Desert climate rivers are found in regions with 200 mm and less of annual rainfall. Shrubs, desert grass or sand cover the basin surface. The rivers are of the wadi type. The drainage network is very poor and if traceable at all, was very likely developed before the present stage of aridity. Dessication has resulted in the obliteration of the lower part of the network and a change from an intermittent perhaps even perennial regime to an ephemeral one. Thus the present stage of the network represents an adjustment to the environment.

4 Rivers of tropical mountains are more extensive in tropical South America and each river is a product of very special conditions. The behaviour of the mountain streams is very similar to that of the rivers of the surrounding ecological regions, though influenced to a greater or lesser degree by the altitude, slope exposure to the wind and rainfall regime deviations. Some of these rivers are under the influence of snowmelt, other are mainly fed from glaciers. Here again the regime depends on the size of the basin and on the altitude of the snowline. Observation of these rivers is very rare except in Ecuador. The river Alao is given as an example (Fig. 4.1.8, Tab. 4.3.8). The gauging station is at an altitude of 3200 metres.

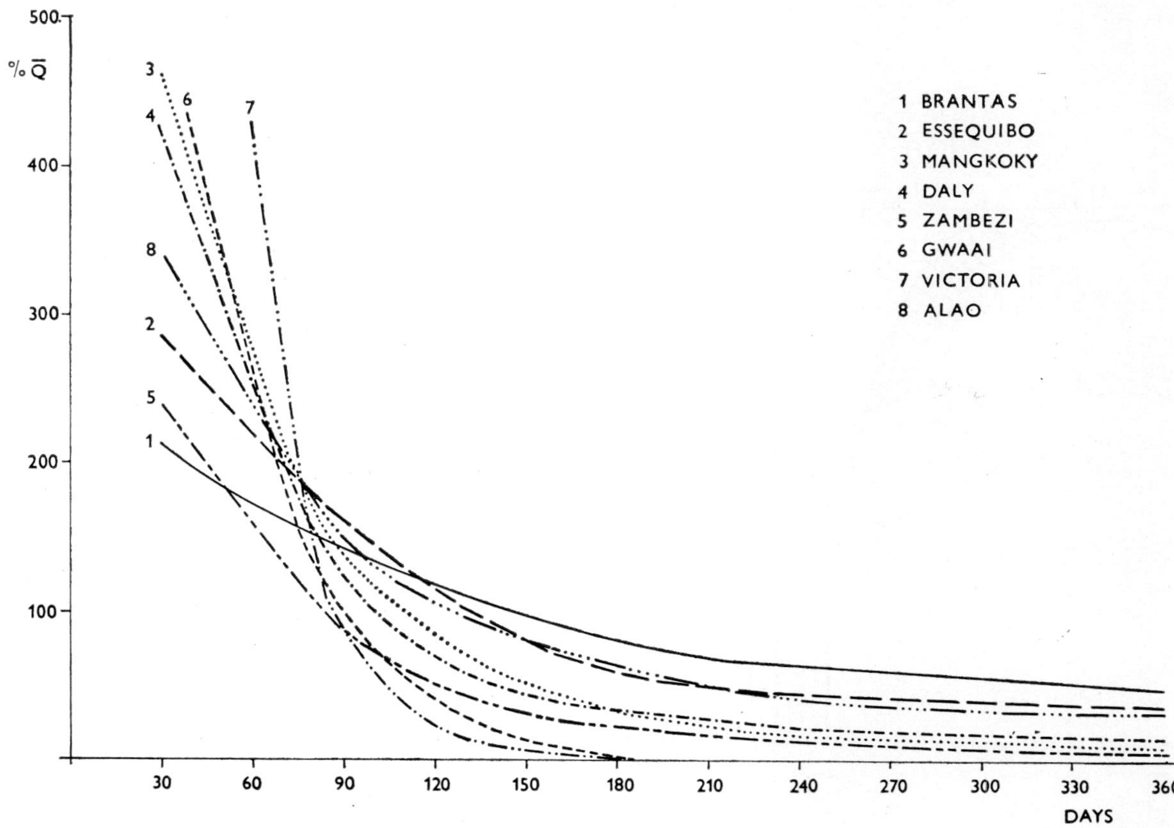

Fig. 4.3 Duration curves for the rivers in Fig. 4.1

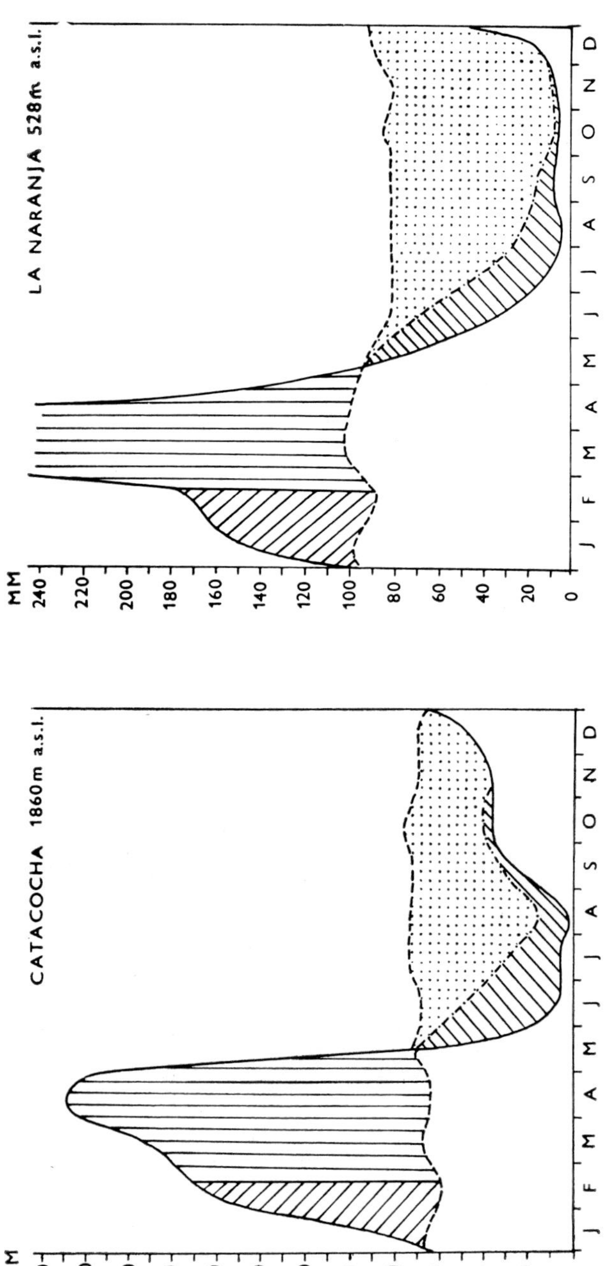

LA NARANJA 528 m a.s.l.
CATACOCHA 1860 m a.s.l.
MM
240
220
200
180
160
140
120
100
80
60
40
20
0
J F M A M J J A S O N D

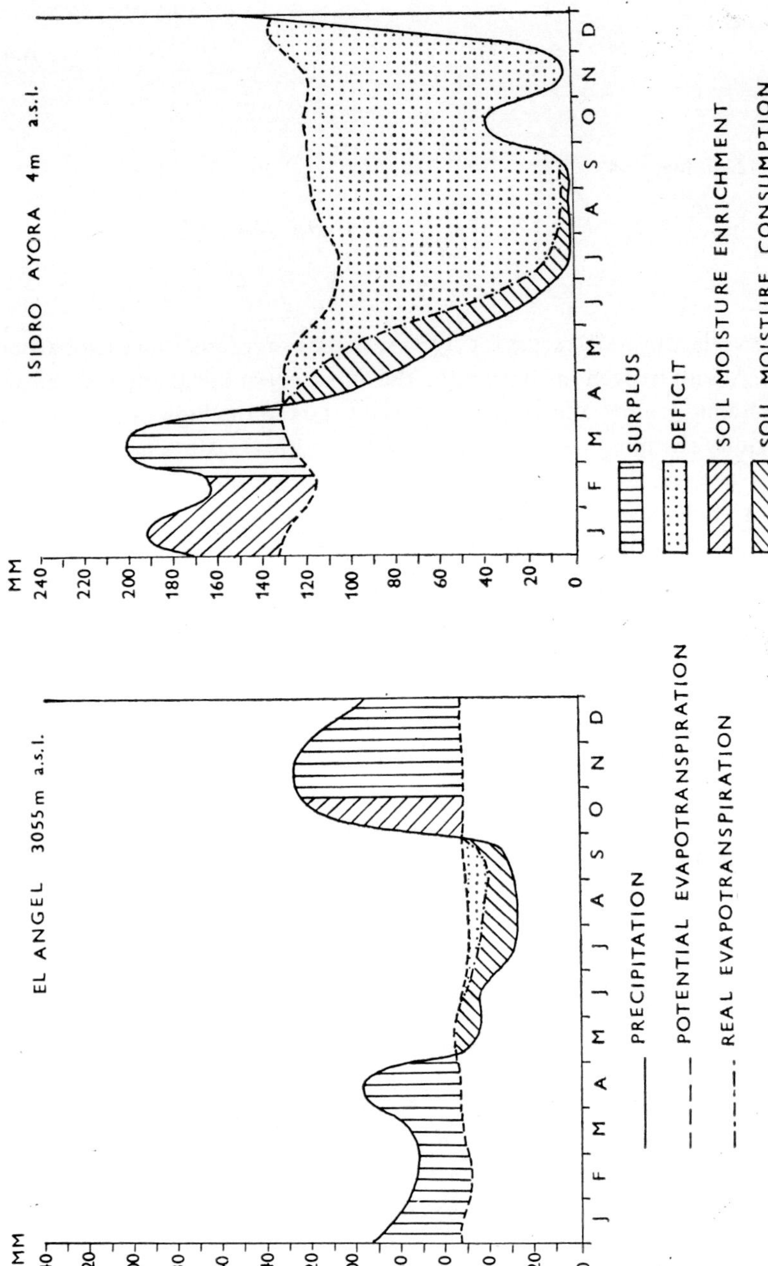

Fig. 4.4 Water balance of four equatorial basins in different altitudes

Tab. 4.3.8 Also basin at En Alao, Ecuador (2°05′ S, 78°25′ W, 130 km²)

	J	F	M	A	M	J	J	A	S	O	N	D	Annual
Rainfall (mm)	122	135	149	175	189	149	54	13	26	95	149	94	1350
Discharge ($m^3 s^{-1}$)	10.5	4.19	4.57	5.39	12.1	25.5	47.2	9.4	23.4	6.5	5.16	11.2	13.8

Fig 4.3 shows the duration curves for rivers selected as representing each particular river regime. As can be seen in the graph, the most economical regimes have the rivers in the humid tropics and rivers with headwaters in the high mountains. With increasing aridity the minimum discharge inclines to lower values or to zero.

Altitude is not the only decisive factor in the formation of the river regime; however, its significance can be studied in regions where the impact of changing altitude can be traced at the same latitude. Hitherto, in Ecuador conditions have been favourable for such a study. Salvador et al. [32] calculated the water balance for almost a hundred cross-sections in Ecuador, most of them close to the equator. Fig. 4.4 gives the monthly balance values for the cross-sections at El Angel, Catacocha, La Naranja and Isidro Ayora located at different altitudes, but having a very similar amount of annual rainfall and approximately the same latitude. The graphical water balance clearly indicates the impact of altitude.

In the world of the tropics some regimes can be found, especially the regimes of small streams, which may not fit into any of the classification schemes and in some way or other are rather peculiar. For instance Alexander [3] observed on the island of Margarita of the Venezuelan coast tropical streams originating from leaf drip supplies of water to numerous springs.

Fig. 4.5 shows the distribution of annual runoff in the tropics.

4.3 GREAT TROPICAL RIVERS AND BASINS

As can be seen in Tab. 4.1 and 4.2, the Amazon is the greatest tropical river by all standards. The hydrological regime of the river and its hydrography have been under intensive study for a long time, although some improvement of existent data can be expected in the future. Two distinctive sections of the longitudinal profile, one steep reaching the confluence with Uyacali and the second flat from Uyacali to the mouth contribute to the unique regime of the river. Below Uyacali the level of the river is only slightly more than 100 metres above sea level, while the headwaters are at an altitude of 6000 metres. Below Uyacali the river is already

1500 metres wide and in lower reaches is over 20 kilometres and at the mouth 80 kilometres.

Typical of the river is a braided channel in a wide shallow valley called várzea. In a cross-section (Fig. 4.6) three different parts can be distinguished. The upper part of the valley is not regularly flooded, in the middle part called terra firme, it becomes a shallow valley regularly flooded by a layer of water several metres deep. The higher areas called campos and the shallow marshlands called igapo are distinctive features of the lowland, through which the main channel meanders. The depth of the main channel is very variable. While in the lower reaches the depth is 6 – 12 metres, upstream depths over 90 metres were found which indicates that the bottom is below sea level.

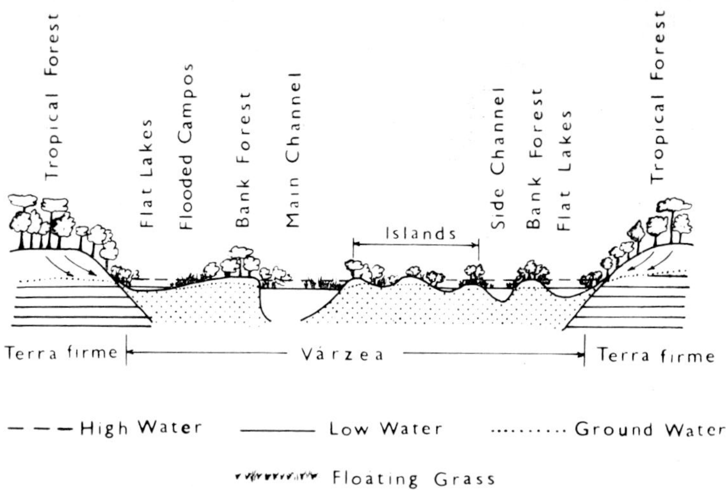

Fig. 4.6 Cross-section of the Amazon valley

The river regime is very regular. The annual hydrograph for a wet year can be seen in Fig. 4.7. More reliable discharge data were not available until 1963. Before that date the discharge in the Amazon delta was estimated to be $80 - 110 \times \times 10^3$ m^3 s^{-1}, with maxima up to 215×10^3 m^3 s^{-1}. After discharge measurements in 1963 new data were established with results indicating that the mean annual discharge is about 212×10^3 m^3 s^{-1}, which corresponds to 6700 km^3 of water per annum. The additional contribution of the River Tocantins, which joins the Amazon in the delta, is 16×10^3 m^3 s^{-1}. Thus the whole Amazon basin contributes 20 % the total runoff from all continents. The water balance is based on the data of Karasik [20], in Tab. 4.4. In the same table the water balance values of other South American rivers are given. The Rivers Tocantins, Madeira and Rio Negro

Tab. 4.4 Water balance of the great basins in South America

River	Drainage area (km²)	Rainfall (mm)	Runoff (mm)	Evapotranspiration (mm)	Yield (l s⁻¹ km⁻¹)	Discharge (m³ s⁻¹)	Runoff coef.
Amazon	6 140 000	2150	1088	1062	35	212 000	0.51
Madeira	3 200 000	1719	543	1176	17	55 000	0.32
Rio Negro	1 029 000	2400	536	1864	17	17 500	0.22
Tocantins	980 000	1555	507	1048	16	15 750[1])	0.33
La Plata	3 104 000	1240	432	808	14	42 400[2])	0.35
Paraguay	1 095 000	1388	234	1154	7	8125	0.17
Uruguay	365 000	1484	522	962	16	6041	0.22
Orinoco	1 050 000	1990	883	1107	28	28 000[3])	0.44
S. Francisco	650 000	1050	183	867	6	3800	0.17

[1]) 26 000 after Netopil [26].
[2]) 42 422 after Keller [21].
[3]) 27 900 after Keller [21].

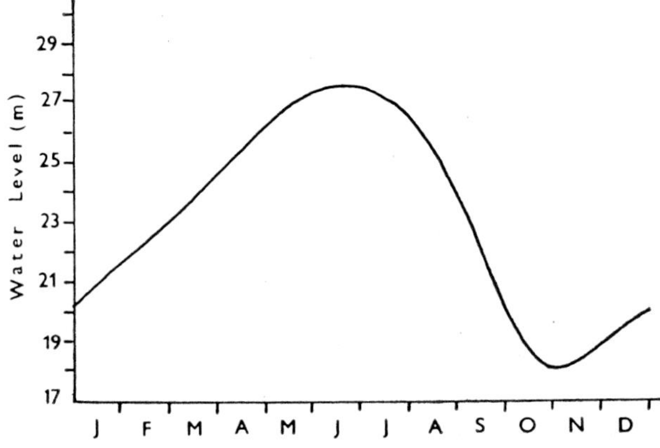

Fig. 4.7 Annual hydrograph of Amazon at Obidos

are tributaries of the Amazon, however, each of them balanced separately belongs to the regiment of the greatest tropical rivers. The Madeira is formed as a confluence of the Rivers Mamoré, Guaporé and Beni. The level of the river rises between October and May as a result of the combined effect of rains on the flat part of the basin and snowmelt in the Andes. The Tocantins joins the Amazon in the delta and is considered by some hydrologists as a separate stream.

A left tributary of the Amazon, the Rio Negro has its upper reaches connected with the basin of the upper Orinoco through the river Casiquiare; however, no information on the water balance of the bifurcation is available.

Another great river system of tropical South America is the Rio de la Plata. The basin does not belong entirely to the tropics, however, the greater part of the headwaters belongs there. The system is developed as a confluence of the Rivers Paraná and Uruguay in the delta below Rosario. Other significant rivers in the basin are the Paranapanema and Iguaçú and Paraguay. There are many rapids and waterfalls in the main stream and tributaries, the most renowned being the 70 metres high waterwall on the Iguaçú. In the estuary of La Plata the level fluctuates not only according to changes in the discharge but also under the effects of cyclonal depressions and wind. A complicated hydrological regime can be seen in Fig. 4.8

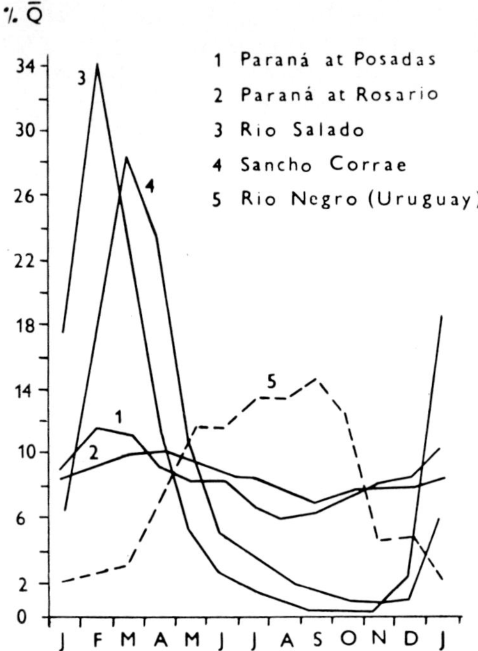

Fig. 4.8 Hydrographs of the La Plata system

where the hydrographs of several rivers in the system are plotted. Basic data are given in Tab. 4.4, together with the data for the tributaries Paraguay and Uruguay.

Another powerful stream in South America is the Orinoco which flows from the Andes parallel to and north of the Amazon. The source of the river was not known prior to 1951. Water balance data are given in Tab. 4.4. The runoff coefficient for the Orinoco is the highest in South America.

The Sao Francisco is also one of the great streams in South America. It drains the western part of Brazil. Owing to a considerably lower rainfall in the basin, the total runoff, yield and the runoff coefficient are also far less than values typical of other rivers.

The regimes of the great rivers of the African tropics are also special in many ways and not entirely relevant to some of the typical regimes given in the classification. Many additional factors make the regime of the great African streams quite unique.

The regime of the Congo is under the influence of combined tropical wet and dry climate and the equatorial wet climate. Fig. 4.9 shows the contribution of several

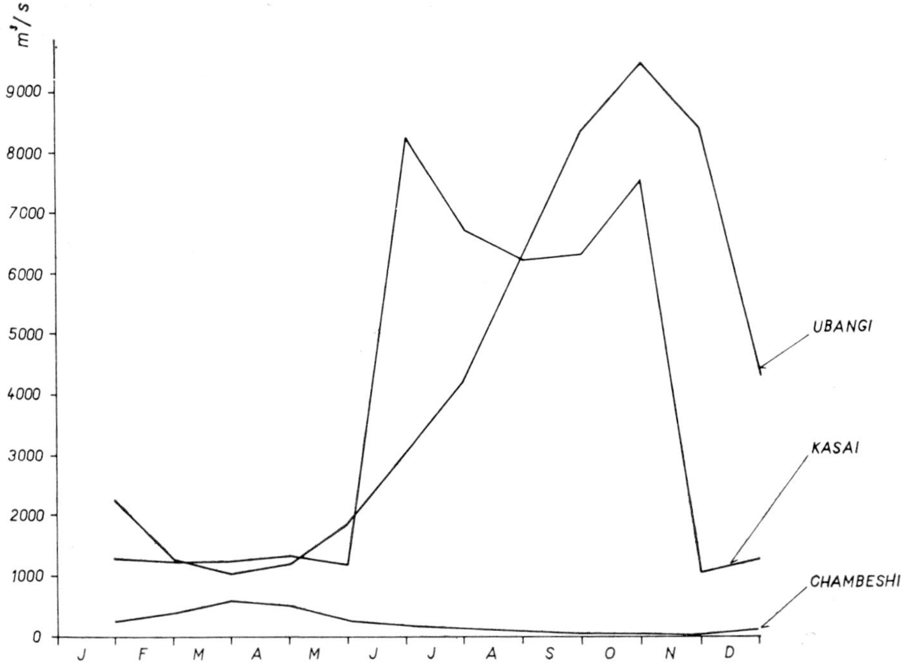

Fig. 4.9 Hydrographs of the Congo system

characteristic rivers, represented by the Chambeshi River in the headwaters, by the Ubangi with one annual peak and by the Kasai with two peaks. The water balance of the river is shown in Tab. 4.5. The regime of the headwaters is transformed in the Bangweulu swamps. Occasionally, even Lake Tanganyika contributes to the river system. In the central parts of the river there are many rapids and waterfalls and the lowest part contains the greatest waterfalls of the world, as far as the amount of water is concerned.

The peak in the headwaters comes some time between March and April. The equatorial tributaries, for instance the Kasai, have the first peak between December and February and the second between March and April, while the Ubangi has a peak flow in September. As a result, the main flood in the lower parts of the Congo comes in November/December and the secondary peak comes in March/April.

Tab. 4.5 Water balance of the great basins in Africa

River	Drainage area (km^2)	Rainfall (mm)	Runoff (mm)	Evapotranspiration (mm)	Yield ($l\ s^{-1}\ km^{-2}$)	Discharge ($m^3\ s^{-1}$)	Runoff coef.
Congo	3 607 450	1561	337	1224	10.7	38 800	0.22
White Nile	1 435 000	710	16	694	0.5	793	0.02
Blue Nile	324 530	1082	158	924	5.0	1727	0.15
Nile	2 881 000	506	28	478	0.9	2590	0.06
Niger	1 091 000	1250	202	1048	6.4	7 000	0.16
Zambezi[1])	1 236 580	759	30	729	1.0	1237	0.11

[1]) At Victoria Falls.

One of the most complicated regimes is that of the Nile. The headwaters are located in the so-called undifferentiated highland south and north of the equator. The river flows through tropical wet and dry regions and in the Sudan it enters immense swamps. Reduced in discharge by the effect of the swamps and joined by the Blue Nile, the river flows through semi-desert and desert regions where another considerable quantity of water is consumed. Artificial reservoir and irrigation schemes have a similar effect. The significance of the contribution of the Blue Nile can be traced in Fig. 4.10. The water balance of the Blue Nile, White Nile and Nile in the delta is given in Tab. 4.5.

The Zambezi basin is under the influence of the tropical wet and dry highland and semi-desert areas of the northern Kalahari. The semi-desert area is more than twice the size of the wet and dry highland. However, as the limited rainfall is lost in the desert, the contribution from the source of the Chobe is rather insignificant. This results in greatly reduced values of the water balance below the mouth of the Chobe flowing from the Kalahari area.

The March flood on the upper Zambezi is delayed by 3–5 months at Lake Kariba. The water balance data are found in Tab. 4.5. Data for the river cross-section at Victoria Falls include the effect of the Kalahari basin. The data for the river mouth, found in Soviet sources, do not treat the Kalahari as part of the basin.

The Niger headwaters are near to the ocean in an area enriched by precipitation on the border between the tropical wet and dry zones. The river then flows into the northern semi-arid region where it loses water through evapotranspiration in the flats above Timbuktu. From there the river flows into the wet and dry tropical lowlands. The contribution of the tributaries can be traced in Fig. 4.11. The water balance data are found in Tab. 4.5.

In Asia only the Mekong and Irrawaddy can be considered to be great rivers influenced by tropical climates. The Mekong originates on the Tibetan Plateau and in its upper reaches it flows through a deep valley and over many rapids. In the lower reaches the valley is flat and is regularly flooded. The flood culminates upstream in August and downstream it is delayed by one month. One of the Mekong branches connects the river with Lake Tonlé Sap where part of the floodwater is stored. Water balance data are given in Tab. 4.6.

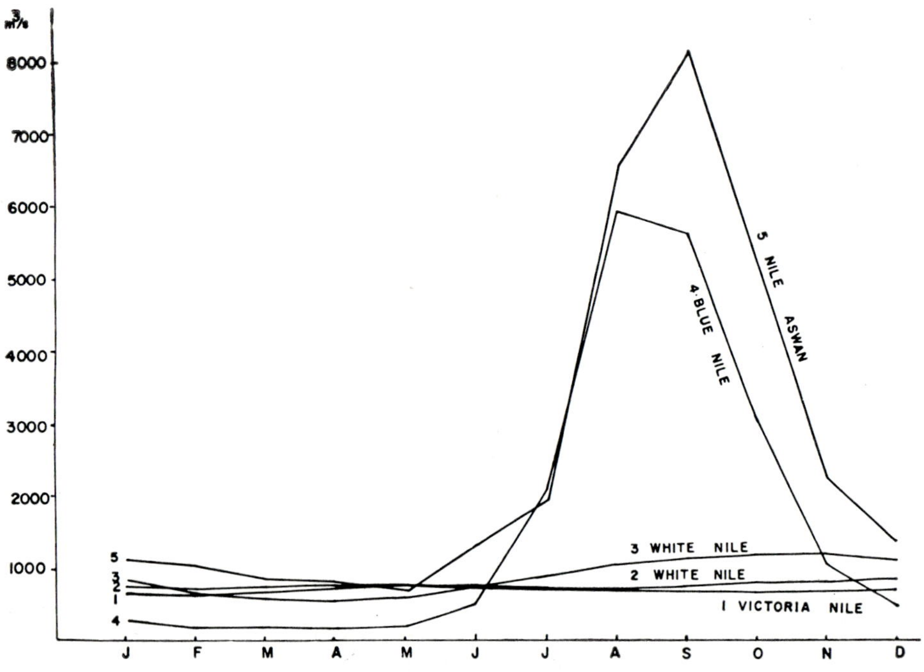

Fig. 4.10 Hydrographs of the Nile system

Tab. 4.6 Water balance of the great tropical basins in Asia

River	Drainage area (km^2)	Rainfall (mm)	Runoff (mm)	Evapotranspiration (mm)	Yield ($l\ s^{-1}\ km^{-2}$)	Discharge ($m^3\ s^{-1}$)	Runoff coef.
Mekong	810 000	1570	523	1047	16.6	13 500	0.33
Irrawaddy	431 000	1970	978	992	31.0	13 400	0.50
Brahmaputra -Ganga	1 730 000	1465	810	655	26.0	44 434	0.55

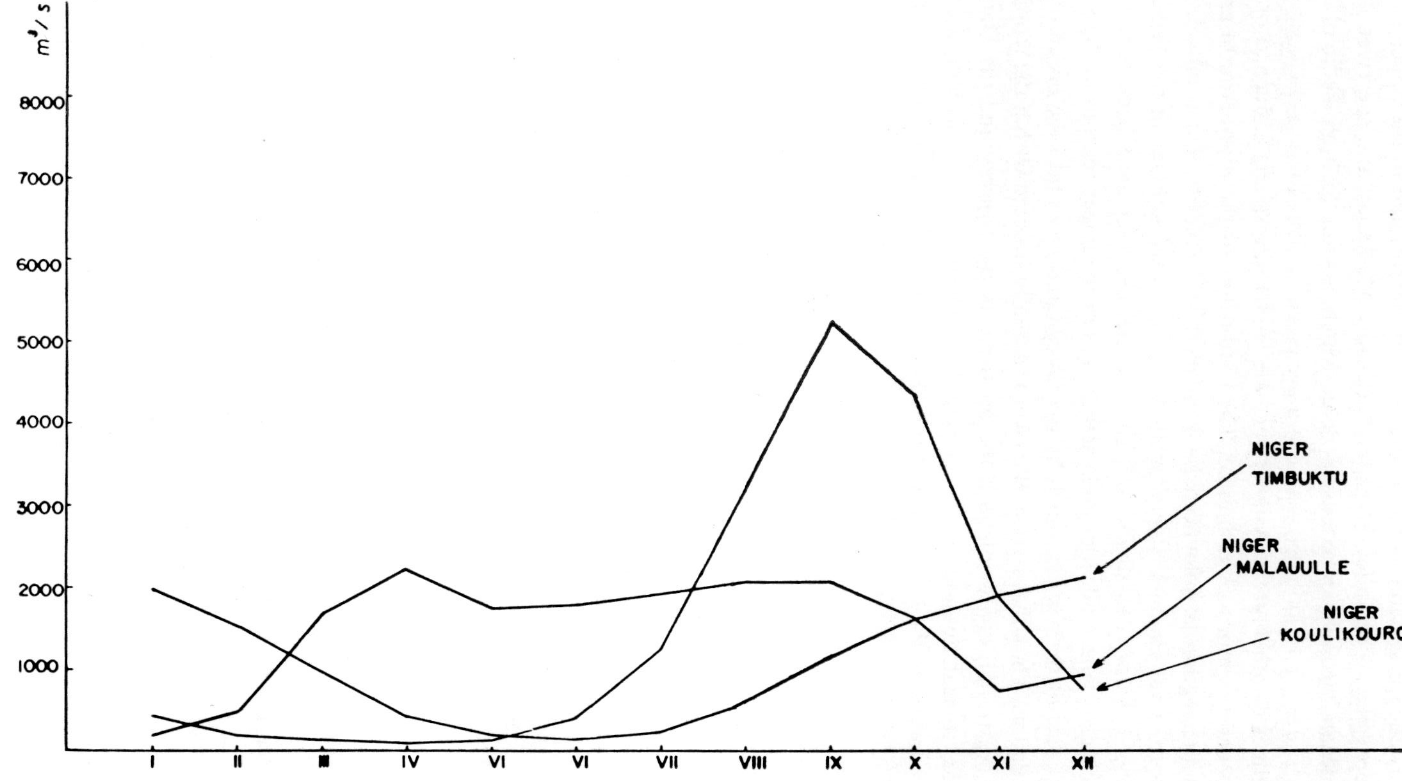

Fig. 4.11 Hydrographs of the Niger system

The Irrawaddy originates at an altitude of 4500 metres and in the upper part it cuts a deep valley. The valley becomes flat below Mandalay and the river flows into the ocean through an extensive delta which, owing to high bedload rate increases its area by 10 km^2 per year. Data are given in Tab. 4.6.

The regimes of other great Asian rivers are under the strong influence of the Himalayas and the monsoonal climate. The headwaters of the Brahmaputra are in Tibet and the regime is influenced by snowmelt and glaciers. Once outside the mountains the river channel is very unstable owing to the amount of the bedload. The channel bottom rises slowly and the channel deposits have been correlated by morphologists with landslides and earthquakes in the upper reaches of the river. The lowest part of the river is joined by the Ganga and flows into the ocean under the name Padma. Only that part of the river is actually situated within the tropics.

On the Australian continent both the great rivers the Murray and the Darling are outside tropical boundaries, nevertheless the Darling is a typical product of semi-arid conditions. Few of its tributaries have headwaters in the vicinity of the tropic of Capricorn.Data are given in Tab. 4.7 as an indication of the regime typical for that part of the continent. A very low mean annual discharge for such a great basin is a typical feature in the water balance.

Tab. 4.7 Water balance of Darling basin, Australia

River	Drainage area (km^2)	Rainfall (mm)	Runoff (mm)	Evapotranspiration (mm)	Yield ($l\ s^{-1}\ km^{-1}$)	Discharge ($m^3\ s^{-1}$)	Runoff coeff.
Darling	590 000	444	13	431	0.6	412	0.03

Tab. 4.8 shows the seasonal variability of the greatest tropical rivers as a percentage of the mean annual discharge.

4.4 LONG-TERM FLUCTUATION OF TROPICAL RIVERS

It has been discussed earlier (Balek [4]) that certain periodical components of river sequences appear to be more pronounced closer to the equator, while with increasing distance from the equator a combination of more or less significant periodicities is found. The strong impact of the sun and also of sunspots on the climate and river regimes has long been the subject of study. Brooks [8] first pointed out that in the tropics such a relationship can be pronounced and his conclusion was later confirmed by Gregory [14]. However, a leading Nile hydrologist Hurst [18] disagreed with this view.

Tab. 4.8 Mean monthly discharge as a percentage of the annual discharge for selected tropical rivers

River	January	February	March	April	May	June	July	August	September	October	November	December
Amazon	65	86	108	120	131	129	129	125	88	71	71	77
La Plata	122	140	137	117	99	97	80	69	71	86	87	95
Congo	108	87	79	171	89	76	72	72	86	103	129	128
Orinoco	50	35	30	38	34	123	160	184	177	141	108	79
Mekong	26	19	15	14	23	87	163	260	285	180	85	43
Irrawady	27	20	21	30	40	99	228	267	175	175	77	40
Niger	28	13	7	4	7	26	84	224	360	297	130	20
Nile	43	40	33	33	28	53	74	250	311	196	86	53
Sao Francisco	118	145	167	150	169	83	56	45	37	43	70	117

Kuzin [23] classified rivers according to their long-term fluctuation into three zones, as can be seen in Fig. 4.12, and found a pattern valid for African rivers.

One of the most regular periodical components is possessed by the Niger, where the periodicity of 25.5 years is well pronounced and has a strong impact on living conditions in the Sahel. Also the periodicities of 7.3 and 3 years were traced.

Periodical components of 21, 7.6, 4.2 and 2.7 years were proved for the Nile, this being based on observations begun in the 19th century.

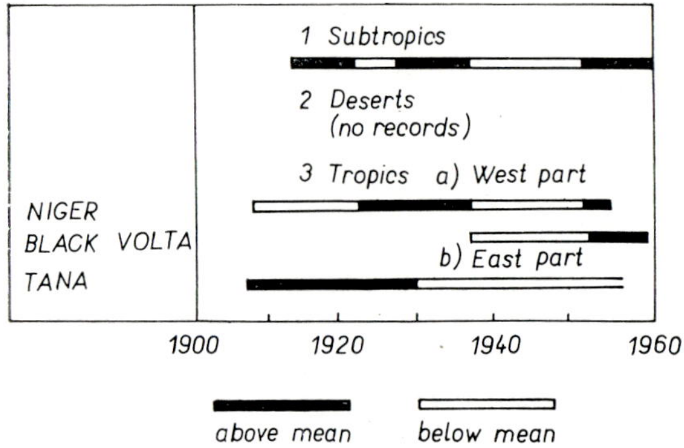

Fig. 4.12 Kuzin's analysis of long-term fluctuation of African rivers

Also the longest hydrological records of maxima and minima, ever found have been analysed for the Nile (Anděl et al. [2]). The Arab historians Taghri Birdi and Al-Hijazi compiled sequences of maxima for 849 years and minima for 663 years, the analysis of which supplied the results given in Tab. 4.9.

Longer periodicities in the table may reflect the rising of the river bed or the existence of even longer periodicities. Remarkably the number seven was found to

Tab. 4.9 Periodicity in Nile historical sequences

Source Sequence of unit	Taghrí Birdi minima (m)	 maxima (m)	Al Hijazi minima (m)	 maxima (m)	Hurst runoff volume (mld m³)
Observed at	Roda	Roda	Roda	Roda	Aswan
Length (years)	663	849	849	849	84
Periodicity	440	556	556	556	20.5—32.2
traced (years)	189—265	242—389	242—389	242—389	7.3
	6.6	151—332	89—113	132—151	
		14—14.2	74—82		
			16.2—18.8		

play a particularly significant role in the shorter periods. Old records are compared with the new short record as compiled by Hurst and the periodicities 7.3 years and 21.7 years of the interval 20.5 – 32.2 traced in Hurst's sequence are related to the periodicities of 77 years (interval 74 – 82 years) as found in Al-Hijazi data and to the periodicities of 6.6 and 14.1 years (interval 14 – 14.2 years) in the Taghri Birdi data.

The periodical analysis of sequences of total runoff from the tropical parts of the continents has remarkable results. These records were compiled [40] for the

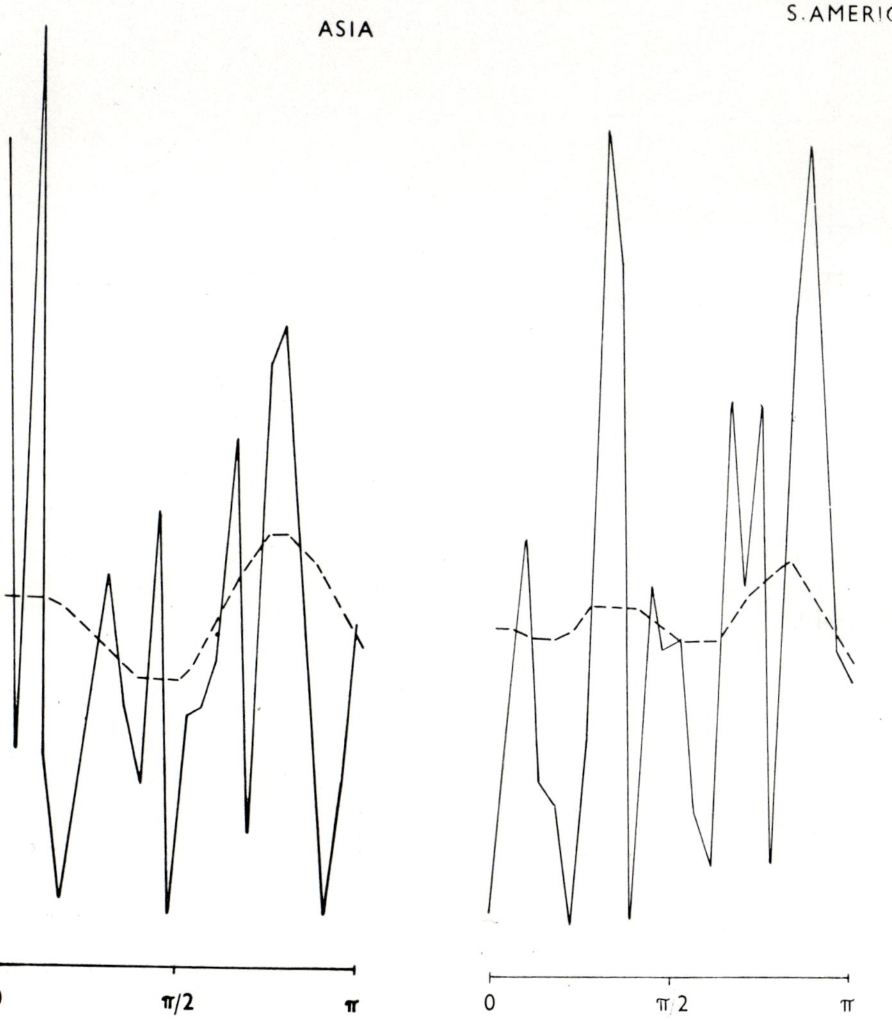

Fig. 4.13.1 Periodograph and graph of spectral density of the annual runoff from the Asian continent

4.13.2 Periodograph and graph of spectral density of the annual runoff from the South American continent

years 1918 – 1967. Periodographs and the graphs of spectral density are given in Fig. 4.13. The periodograph for tropical Asia has a strong periodical component of 2.5 years and a second of 16.6 years. A similar strong component of 2.38 years was found for South America, where the period of 25 years and 6.3 years were traced. For Australia the shorter periodicity found was 3.3 years, however the periodical component of 10 years was much stronger. Finally, for Africa the strong-

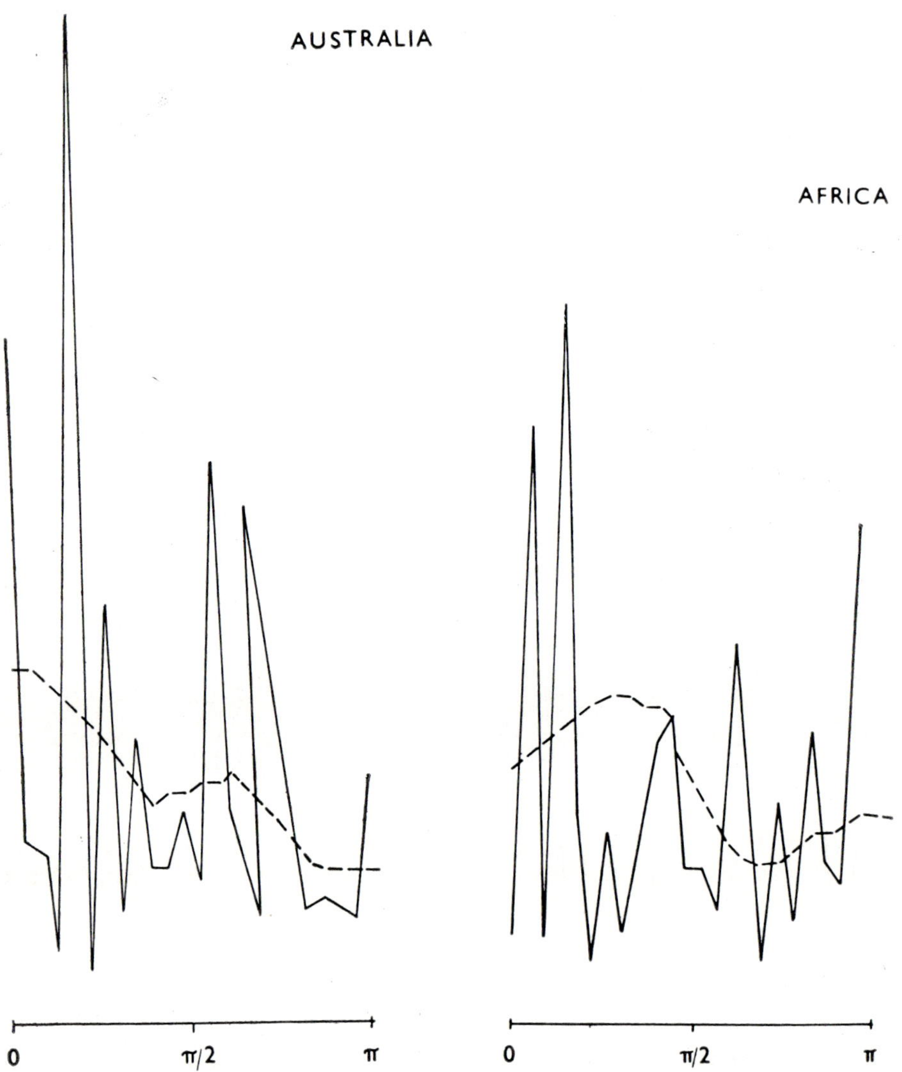

Fig. 4.13.3 Periodograph and graph of spectral density of the annual runoff from the Australian continent

Fig. 4.13.4 Periodograph and graph of spectral density of the annual runoff from the African continent

est component is 5.6 years followed by another component of 12.5 years. The relative shortness of the sequences does not allow any more definite conclusions, perhaps with the exception of short periods of 2.3 – 5.6 years for the occurrence of which some common reason can be sought.

However, long-term forecasting does not go beyond approximate estimates. There is still some scepticism about the existence of relationships between hydrological and hydrometeorological sequences on the one hand, and other phenomena which have been proved experimentally such as sunspots on the other, (Anděl, Balek [1]).

4.5 SOIL EROSION IN TROPICAL BASINS

Erosional problems in the tropical world are in many ways more serious than in regions of moderate climate. It has been calculated that the rivers of Java carry more silt in one day than European rivers of the same drainage area in two centuries. The river Irrawaddy itself carries an average of 1 kg of silt per m^3 of water. However, no systematic studies on erosional processes for the tropical regions have been reported so far. According to Starmans [36] who analysed erosional processes in Asia and Africa the following factors are the most decisive in tropical erosional processes:

a) The aridity and/or semi-aridity of climate,
b) the degree of destruction of vegetation,
c) the ability of vegetation to regenerate,
d) the leaf area index of the vegetation,
e) the state of soil protection against violent temperature changes,
f) the age of river basins,
g) the regime of rainfall.

A relationship between the percentage of a basin adequately covered by vegetation and silt load has been developed by Starmans [36]. He concluded that hilly tropical areas well vegetated, as in parts of Africa, produce quite high floods of short duration and the mitigating effect of the vegetational cover on the erosional processes is evident even on a large scale. It is worth noting that the amount of dissolved solids exceeds the amount of suspended solids in some basins.

The positive influence of vegetation on the stability of the basin is also indicated by results obtained by Balek and Perry [5] indicating that the annual amount of soil washed from areas up to 90 – 95 % covered by tropical highland forest was only between 0.000053 and 0.000142 mm per annum.

In savanna gullies erosion is uncommon, except in overgrazed areas of the drier parts of the savanna. In general red soils are more resistant to erosion.

Bostanoglou [7] listed highly efficient methods to protect watersheds against erosion in semi-arid and arid regions. The most suitable method is the planting of suitable vegetational cover shadowing the soil and stabilizing it by the root system. For arid regions *Ceratonia siliqua, Acacia* and *Gleditschia* are recommended. To improve the physical and chemical properties of the soil the following plants have been found suitable: *Salix, Hippophas, Acacia, Alnus, Pinus nigra, Robinia pseudoacacia, Populus, Acer, Sambucus, Elaeagnus, Ailanthus, Betula* and *Rhamnus.*

Only some of a wide selection of species may be found to be suitable from an ecological point of view. Balek [6] recommended the selection of species which can survive with the lowest rate of water consumption. A list of a wide variety of species most suitable for particular tropical countries has been published by Harcharik and Kunkle [15]. For example *Acacia auriculaeformia* and *Acacia glausa* are recommended for Thailand and Indonesia, *Alnus maritima* for Indonesia and various types of bamboo for all regions where other plants fail to survive.

It is always recommended that certain species should be stabilized on a small scale and the scheme only extended if successful. For instance Stevens [37] reported that within eight years of establishing the fast growing *Tamarix aphylla* on the sands surrounding an oasis in Saudi Arabia the depth of the organic horizon increased from zero to 1 mm and the carbonate content decreased from about 30 to 15 %.

The planting of grasses and the formation of small terraces is another mean of protection. To stabilize the grass carpet pre-planted at other locations and laid on the soil, stabilizing pegs have to be applied. On unconsolidated soils pegs of fascines can be used as a basis for the construction of small terraces.

The stabilization of mountain streams in the tropics was widely analysed by Hattinger [16]. It can be assumed that the laws of channel hydraulics are the same for tropical regions as for moderate ones. However, as stated by Hattinger, in the tropics increased channel erosion is most frequent in deforested regions or where road construction has resulted in the uncontrolled deviation of existent river channels and gully erosion has occurred on denuded areas. Thus rapid changes of the hydraulic properties of the channels and an increased flood rate affect the erosional processes very rapidly.

As reported by Tejwani [38] from India a survey of 21 reservoirs indicates that they receive 85 mm of sediment from the basin, although the design value was only 30 mm. Grove [13] published the erosion rates for west African streams which are given in Tab. 4.10. They are far below possible maxima. A comparison of soil loss for tropical and temperate regions was made by Kirkby and Morgan [22] in Tab. 4.11.

In general increased soil loss in the tropics was accounted for by higher rainfall rates of tropical storms. Dunne [10] observed erosion ranging from between 20 and 200 tons per km^2 per year in various parts of the tropics, this being dependent on the climate. He attributed most of it to the construction of country roads and the impact of changing wet and dry periods.

Tab. 4.10 Erosional rates for some rivers in West Africa

River	Niger	Benue	Niger-Benue	Congo
Discharge (10^{-12} year^{-1})	0.1	0.11	0.21	1.2
Erosional rates				
(10^{-6} tons year^{-1}) in solution	5.5	4.5	10.0	98.5
suspended	9	22	31	31.2
combined	14.5	26.5	41	129.7
(tons year^{-1} km^{-2})	19	77	37	37
(10^3 mm year^{-1})	8	36	15	14.8

Tab. 4.11 Soil loss (kg m^{-2} year^{-1}), in the tropics and temperate regions (after Kirkby and Morgan)

Plant	Slope	Tropics	Temperate regions
Forest	—	.009 rainforest	0.003
Grass	4—9	.005—.02	0.007
Bare soil	4—19	10—17 humid tropics	1.0
		1.8—3 savanna	
Maize	—	.03	12.8
Coffee	—	2.2	
Banana	0—7	1.5	
Manioc	0—7	9.0	
Sorghum	2	0.3—1.2	
Groundnuts	2	0.3—1.2	
Rice	—	0.017—0.289	

Hudson [17] reported a significant reduction in soil loss in Zimbabwe from 1.23 to 0.07 kg m^{-2} year^{-1}, after the density of maize planting was increased from 2.5 to 3.7 plants per m^2 and trash mulch applied.

Elwell and Stocking [11] developed a rational formula for the estimation of soil loss, based on experiments in South Africa:

$$Z = KCX$$

Here Z is the predicted mean annual loss, K is the mean annual loss from a standard field plot of 30 × 10 m at a 4.5 % slope for a soil of known erodibility under bare fallow, C is the ratio of soil lost from a cropped plot to that lost from the standard plot, X is the ratio of soil lost from a plot of length L and slope S to that lost from the standard plot.

One of the most widely used methods for the prediction of soil loss is the Universal Soil Loss Equation (USLE), in the form:

$$A = 0.224\ RKLCSP$$

in which A is soil loss (kg m^{-2}), R is the rainfall erosivity factor, K is the soil erodibility factor, L is the slope length factor, S is the slope gradient factor, C is the cropping management factor, P is the erosion control practice factor. Data for the assessment of these factors are available in various handbooks and are applicable even in the tropics with the exception of the rainfall erosivity factor developed by Wischmeier. This is actually a definition of the erosivity of single rainfall events and is thus defined as a product of two rainstorm characteristics, namely kinetic energy and the maximum 30 minute rainfall rate. For Hawaii it was recommended to use for R

$$R = 0.0134P^{2.2}$$

where P is 6-hour duration rainfall (mm), for a 2-year recurrence interval. Similarly, for west Africa Roose [31] found

$$R = H(0.50 \pm 0.05)$$

where H is the average amount of rainfall, mm.

Many attempts have been made to set up critical values for the occurrence of the erosional processes. Critical rainfall for Zimbabwe was determined by Hudson as 25 mm h^{-1}. Some other authors believe that the value is only valid for tropical areas with an intense rainfall pattern. A critical value for the occurrence of wind erosion was found to be 4 m s^{-1}. A rate of 11 tons per hectare per annum is generally considered as an acceptable maximum soil loss.

For the terracing design several formulae have been developed in the tropics. For the then southern Rhodesia it was found to be

$$V = 0.61(S + f)$$

where f varies from 3 to 6 according to the variability of the soil, V is the vertical interval between terraces (m), and S is the slope of the terraces (%). For South Africa

$$V = .3\left(\frac{S}{a} + b\right)$$

where a varies from 1.5 for low rainfall to 4 for high rainfall and b varies from 1 to 3 according to the type of soil. For Kenya

$$V = 0.3\frac{S+2}{4}$$

So far no general formula for the tropics is available.

In taking protective measures agricultural land is differentiated from land not suitable for agriculture. The cheapest means for the protection of agricultural lands is contour cultivation which reduces the surface runoff, prevents soil erosion and conserves soil fertility. As a second stage contour bunding, graded bunding and bench terracing on steep slopes is used. A contour bund system is a protective measure recommended in semi-arid lands with high infiltration and permeability. Bunds are commonly adapted to a slope of about 6 %. Alternatively especially for deep black soils contour ditches are used. Graded bunds are used in areas having more than 800 mm of rainfall per annum, irrespective of soil texture. In clay soils graded bunds are used for areas with less than 800 mm of annual rainfall.

On steeply sloping and undulating land bench terracing is recommended. Although the initial cost is high, the system is more productive and makes the application of sprinklers, fertilizers etc. possible. In rainfed areas it is used on slopes from 6 to 33 %.

Runoff store techniques vary from place to place and depend on the budget available. In monsoonal areas excess water can be stored in ponds or underground through artificial infiltration. The cost of lining large ponds can be too expensive and is not suitable everywhere.

Land not suitable for agriculture may be characterized by steep slopes, severe erosional processes, rockiness, shallow soil profile, flooding, arid climate, etc. However such land may have a great potential for producing fodder, fuel fibres, minor fruits and low quality timber. Improvement is achieved by protecting the natural vegetation and improving the quality of grasses. *Eucalyptus hybrid* for fuel and *Eulaliopsis* for fodder and industrial grass for rope and paper can be produced.

A remarkable method for fighting erosional processes was reported by Buck [9]. He developed a system of domesticating soil bacteria which were capable of producing a great number of polyuronic volloids. These were extremely effective in increasing those properties of the soil which help to resist soil erosion. Through this method he actually transformed 53,000 hectares of southern Gran Chaco eroded bushland into good pasture.

The reduction of surface runoff is another means of erosional protection, usually performed on a large scale. Jorge [19] conducted several experiments in Cuba on the formation of forest belts along the rivers. Narrow strips of forest were capable of absorbing the rainfall over the forest as well as the runoff from the upper parts of the basin. The method was applied in the basin of San Diego River.

In arid regions straw mulches and pebbles were widely applied to conserve soil moisture and reduce wind erosion and this also suppressed weed growth. Recently the application of plastic mulches has been practised, although grass, hay, straw etc. are much cheaper. By using coloured types of mulches the soil temperature was regulated.

For sediment sampling in remote cross-sections of tropical rivers special devices consisting of several single stage samplers have been developed. The device consists of several samplers arranged on a staff so that a sample is taken automatically with each rise of the water level. The device was developed by Gilmour [12].

A large scale measurement of the load carried by the West African rivers was conducted during the Trans-African Hovercraft Expedition (Grove [13]). There are many additional features such as loss of water downstream, evaporation from the water surface etc., for which the results can be considered as only indicative (Fig. 4.14), especially when based on a chemical analysis. Hurst provided very useful data on dissolved solids in various cross-sections in the Nile basin (Tab. 4.12). It can be assumed that seasonal variations from given values are rather high, from the results of measurements on the Nile at Wadi Halfa, reported by Sinaika [34]. His measurements are found in Fig. 4.15.

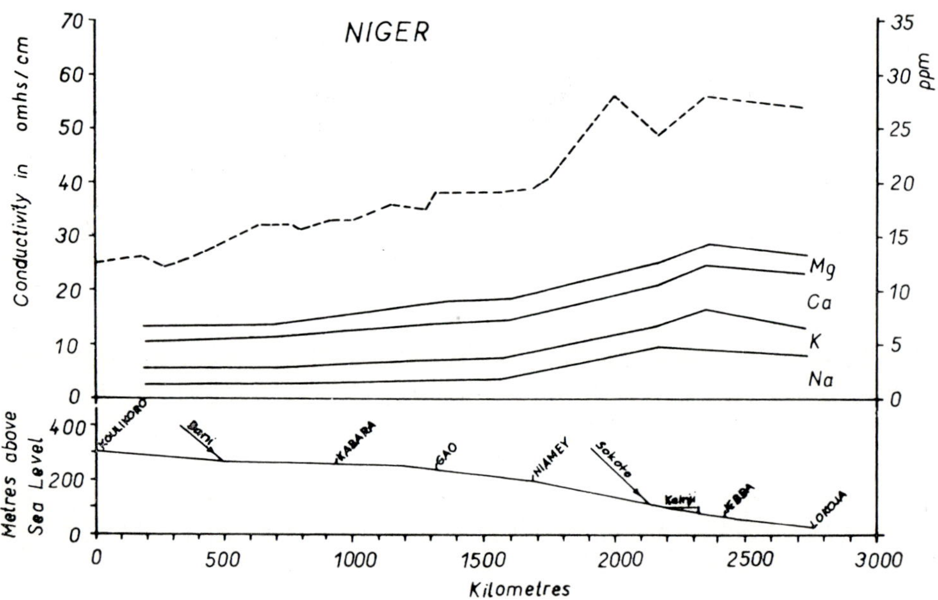

Fig. 4.14 Dissolved load of Niger according to Grove

Pickup and Higgins [29] attempted to find the best formula for the calculation of sediment transport in braided gravel channels of tropical rivers. They experimented on the Kawerong river, Papua New Guinea, and found the Meyer–Peter equation the most suitable.

Tab. 4.12 Dissolved solids, by weight in various cross-sections of Nile basin

Lake Victoria	80 ppm
Victoria Nile below Kyoga	100 ppm
L. Edward	670 ppm
L. Mobutu	590 ppm
Lake Tana	170 ppm
Albert Nile below Mobutu Lake	160 ppm
Blue Nile at Khartoum	140 ppm
White Nile at Khartoum	130 ppm
Atbara	200 ppm
Nile at Cairo	170 ppm

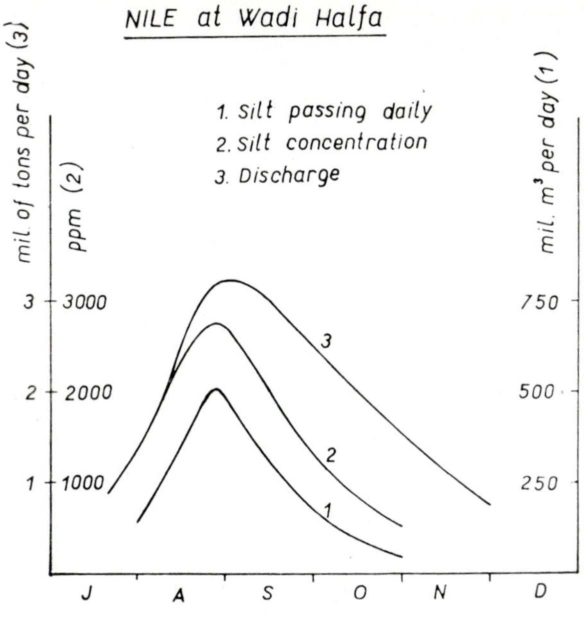

Fig. 4.15 Nile silt observation according to Sinaika

$$\frac{q^{2/3}S}{D} = 17 + 0.40\frac{q_s^{2/3}}{D}$$

where q is the flow discharge per unit of time and width, (kg m^{-1} s^{-1}), S is the energy slope fraction, D is the representative grain size (m) and q_s is the rate of sediment load per unit of time and width (kg m^{-1}s^{-1}). For a mixture D is grain size of such a fraction that 35 % of the load material is finer.

4.6 REFERENCES

[1] Anděl, J., Balek, J., 1971. Analysis of periodicity in hydrological regimes. Journal of Hydrology 14, 66—82.

[2] Anděl, J., Balek, J., Verner, M., 1971. An analysis of historical sequences of the Nile maxima and minima. Symp. on the role of hydrol. in the econom. dev. of Africa. WMO Rep. No. 301, Addis Ababa, 15 p.

[3] Alexander, C. S., 1958. The geography of Margarita and adjacent islands, Venezuela. Univ. Calif. Geol. Publ. 12, 85—192.

[4] Balek, J., 1972. Tropical hydrology. In Proc. of Univ. Sem. Vol. 5, Ed. G. Halasi-Kun. Columbia Univ., 64—113.

[5] Balek, J., Perry, J., 1972. Luano catchments, first phase final report. NCSR WR 28 Rep., Luska, 120 p.

[6] Balek, J., 1978. Engineering the balance between man, land and water. In "In the Spirit of Enterprise from Rolex Awards", Ed. G. B. Stone, Freeman, San Francisco, 159—162.

[7] Bostanoglou, L., 1980. Restauration and protection des pentes degradees. In "Conservation des Resources Naturelles en Zones Arides". FAO Rome, 135 p.

[8] Brooks, C. E. P., 1923. Variations in levels of the Central African lakes. Geogr. Mem. No. 20.

[9] Buck, J. S. M., 1981. Domesticating soil bacteria to fight erosion and create energy. Rolex Awards for Enterprise, Geneva.

[10] Dunne, T., 1979. Sediment yield and land use in tropical catchments. Journal of Hydrology 42, 281—300.

[11] Elwell, H. A., Stocking, M., A., 1976. Vegetational cover to estimate soil erosion in Rhodesia. Geoderma 15, 61—70.

[12] Gilmour, D. A., 1976. Sampling streams for suspended sediment, with reference to tropical rainforest observations. In "Hydrol. Techniques for Upstream Conservation". FAO Rome 1976, 133 p.

[13] Grove, A. T., 1971. The dissolved and solid load carried by some West Afr. rivers. Journal of Hydrology 16, 287—300.

[14] Gregory, R., 1930. Weather recurrence and weather cycles. Quat. Journal Royal Met. Soc. 193.

[15] Harcharik, D. A., Kunkle, S., H., 1978. Forest plantation for rehabilitating eroded lands. In "Special Readings in Conservation", FAO Rome, 101 p.

[16] Hattinger, H., 1978. Correction des torrents de montagne, notamment sous les tropiques. In "Techniques Hydrologiques de Conserv. des Terres et de Eaux en Montagne. FAO Rome, 135 p.

[17] Hudson, N. W., 1980. Soil conservation. Betsford, London, 320 p.

[18] Hurst, H. E., 1954. Le Nil. Payot, Paris.

[19] Jorge, R. P., 1975. Forestry influence in the quality of water. I.N.D.A.F. Rep., Cuba, 12 p.

[20] Karasik, G. J., 1974. Water balance of South America. (In Russian). Sovietskoe Radio, Moscow, 110 p.

[21] Keller, R., 1952. Gewässer and Wasserhaushalt des Festlandes. G.B. Teubnervelagsgesellschaft Leipzig, 250 p.

[22] Kirkby, M. J., Morgan, R. P. C., 1980. Soil erosion. J. Wiley, New York, 312 p.

[23] Kuzin, N., 1970. Cyclic variations of river runoff in Northern Hemisphere. (In Russian). Gidrometeoizdat Leningrad.

[24] Ledger, D. C., 1975. The water balance of an exceptionally wet catchment area in West Africa. Journal of Hydrology, 24, 207—214.

[25] Morgan, R. P. C., 1980. Implications. In "Soil Erosion", Ed. M. J. Kirkby, R. P. C. Morgan. J. Willey, New York.

[26] Netopil, R., 1972. Hydrology of the continents. (In Czech). Academia, Prague, 294 p.
[27] Obi, M. E., Asiegbu, B. O., 1980. The physical properties of some eroded soils of southeastern Nigeria. Soil Sc., Vol. 130, No. 1, 39—49.
[28] Pardé, M., 1953. Fleuves et riviéres. Armand Colin, Paris, 224 p.
[29] Pickup, G., Higgins, R., 1979. Estimating sediment transport in a braided gravel channel. Journal of Hydrology, 40, 283—298.
[30] Daniel, J. R. K., 1981. Drainage density as an index of climatic geomorphology. Journal of Hydrology 50, 147—154.
[31] Roose, E. J., 1971. Influence des modifications du milieu sur l'erosion. ORSTOM, Ivory Coast, 22 p.
[32] Salvador, H. C., Cruz, L. A., Lozada, G., Gómet, E., 1974. Balance hidrico de localidades ecuatorians. Inst. Nat. de Met. e Hidrol. Publ. No. 14, Quito, 101 p.
[33] Sanchez, P. C., 1923. Estudio hidrologico de la Republica Mexicana. Publ. No. 17, Sec. de Agric. y Fomento, Mexico.
[34] Sinaika, Y. M., 1940. The suspended matter in the Nile. Cairo Schindler's Press.
[35] Springfield, H. W., 1978. Using mulches to establish woody chenopods-an arid land example. In "Special Readings in Conservation", FAO Rome, 101 p.
[36] Starmans, G. A. N., 1970. Soil erosion of selected African and Asian catchments. Int. Symp. on Water Erosion, Prague, 10 p.
[37] Stevens, J. H., 1974. Stabilisation of aeolian sands in Saudi Arabia's Al Hasa oasis. J. Soil and W. Cons. 29, 129—133.
[38] Tejwani, K. G., 1979. Malady-remedy-analysis for soil and water conservation in India. Ind. Journal of Soil Cons., Vol. 7, No. 1, 29—45.
[39] Voeykov, A., 1884. Flüsse und Landseen als Produkte des Klimas. St. Petersburg.
[40] Anon., 1974. World water balance and water resources of the earth. (In Russian). Gidrometeoizdat, Leningrad, 637 p.

5 GROUNDWATER AND WATER IN SOIL

5.1 GENERAL

In tropical regions, especially in their semi-arid, arid and wet and dry parts, groundwater is one of the most important natural resources. Even im humid parts of the tropics groundwater plays an important role since many surface water sources are intermittent or polluted and thus domestic supply, cattle watering, agriculture and in some cases industry, depend on groundwater resources.

Surface waters are not always fit to drink because of the presence of various bacteria, parasites and contamination. Even in the humid tropics surface water resources are characterized by scarcity, uneven distribution inadequate development, or overwhelming abundance. Thus the establishment of a low cost water supply serving social and economic needs without endangering the environment can be achieved through the utilisation of groundwater resources.

A systematic survey of tropical groundwater resources can be characterized as spontaneous, even if various projects of U.N.D.P. have been conducted throughout Latin America, Africa and Asia to serve the needs of large populations and thousands of wells have been drilled all over the tropical world. At present priority is given to projects for the supply of water for drinking and domestic purposes, second priority is given to water supply for cattle and herds, followed by supplies for irrigation purposes, industry, and for the development of tourism.

Sometimes, however, groundwater utilisation and the drilling of boreholes may become an integrated part of a special project or of emergency measures, as happened during an extensive drought in Cameroon, where groundwater supply for pepole and herds had to be provided at the same time.

The present trend in groundwater exploration differentiates between renewable and non-renewable resources and leads to the determination of maximum and minimum water levels in order to regulate storage capacity. The effect of man's activities should be borne in mind in order to prevent unfavourable economic conditions which would result in damage to aquifers. Such damage can be irreversible. Reports from various parts of the tropics of a steady decrease in groundwater level are alarming. A decrease in the groundwater level due to excessive pumping, beside causing a permanent damage to the stability of groundwater storage, increases the cost of further pumping and facilitates salt water migration and the intrusion of seawater into the coastal aquifers.

Protective measures in regions with a steadily decreasing groundwater level are:

a) redistribution of pumping schemes so that the rate of the drawdown can be reduced,

b) increase in groundwater recharge through artificial infiltration,

c) importation of water from less exploited regions,

d) increase of recharge through the improvement of the microclimate and physical properties of the soil,

e) limitation of pumping to the rate of natural recharge,

f) legal and administrative protection of aquifers.

5.2 GROUNDWATER STORAGE, ITS DEPLETION AND REPLENISHMENT

The process of groundwater storage enrichment and depletion has to be seen as an integral part of complex interactions between the atmosphere, vegetation, soil and river network. Apart from the water stored in aquifers during past geologic periods the following are the sources of groundwater recharge:

a) condensation of atmospheric humidity,

b) rainfall,

c) infiltration of surface water from streams and lakes.

The first source can be regarded as insignificant in the tropics. The recharge from rainfall is significant where rainfall coincides with increased humidity, otherwise the recharge is reduced partially or totally by evaporation and transpiration. An important role is played by rock formation. Denuded bare rock containing fissures significantly increases the recharge, whereas vegetation reduces it either directly or transpiring the stored groundwater. In particular mention should be made of the impact of the phreatophytic plants; they are able to consume an enormous amount of water without providing any economic value.

Intermittent streams are a significant source of groundwater replenishment which can exceed the contribution of rainfall when river beds are sandy or with fissures. A significant part of the drinking water in Gwembe Valley above Kariba reservoir comes from the groundwater accumulated in the river beds of intermittent streams. And in arid regions groundwater replenishment from temporary river channels can significantly exceed the contribution of rainfall in limited but appreciably significant zones.

The replenishment of groundwater in arid regions was studied by Martin [29] in the Kalahari. He considered that the best areas available for replenishment are those with a uniform water table between 6 – 30 metres and a sand layer not exceeding 0.3 – 1.2 metres. With the increase of the sand layer, say to over 10 metres the groundwater level shows very irregular behaviour. A sand layer is believed to provide good protection against water loss through evaporation. Extensive experiments on evaporation loss from the potential groundwater replenishment were performed in south Africa. Hellwig [19] proved that a fall of the groundwater table below 60 cm in sand with a mean diameter of 0.52 mm will practically prevent evaporation losses. In Kalahari he measured evaporation from the free surface and found it higher by 8 % than from saturated sand. A series of results in Fig. 5.1 indicates the dependence of daily evaporation on the depth of the groundwater table. Also daily fluctuation of evaporation indicates a high degree of variability of evaporation even for a constant groundwater level (Fig. 5.2). Later Hellwig

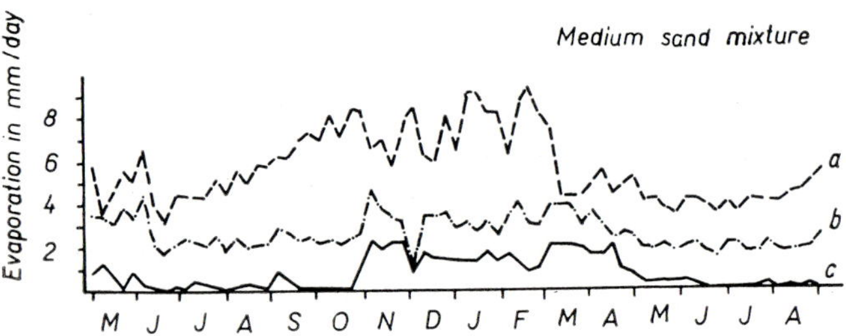

Fig. 5.1 Evaporation from sand as a function of the depth of the water table, according to Hellwig a, water table at the sand surface, b, 30 cm below sand surface, c, 60 cm below sand surface

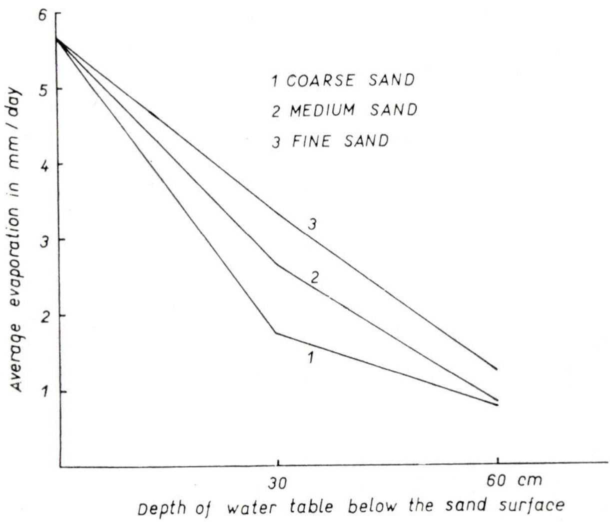

Fig. 5.2 Fluctuation of evaporation from sand as a function of the depth of the water table in medium sand mixture, according to Hellwig. a, water table at sand surface; b, 30 cm below sand surface; c, 60 cm below sand surface

[20] proved that there were two peaks in diurnal fluctuation of water evaporation from sand. One peak he related to radiation and he found its timing to be dependent on the depth of the groundwater table. He found the second peak at sunrise and related it to the air temperature and to the condensation of water at the surface.

A significant amount of groundwater can be consumed by transpiration. The roots of tropical forest species tend to establish themselves in such a way that they are always in a contact with the groundwater storage through the zone of the capillary rise so that they can tap it when access to water in the non-saturated soil levels becomes difficult. Fig. 5.3 shows soil moisture fluctuation in poorly drained and slowly permeable soil under woodland on the Central African Plateau. The groundwater level here was at a depth of about 6 metres and the fluctuation of soil

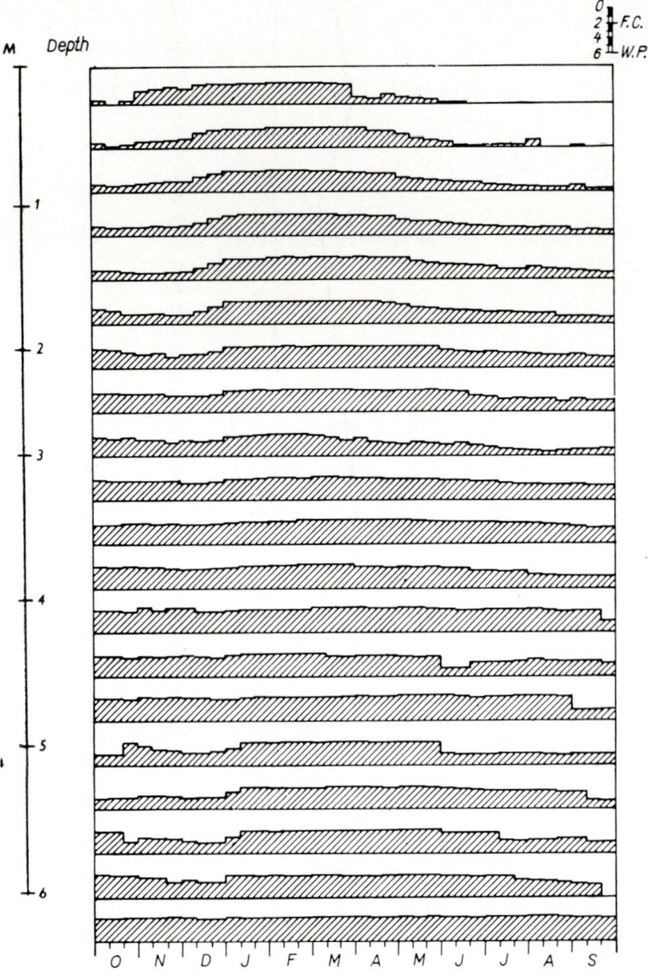

Fig. 5.3 Soil moisture fluctuation in poorly drained slowly permeable soils under woodland on the Central African Plateau. Water uptake from unsaturated and saturated zone can be recognized

moisture clearly indicated increasing consumption of groundwater through the capillary rise particularly during the dry season. At this location almost all annual rainfall – well over 1200 mm – was consumed by evapotranspiration and a significant part of precipitation was temporarily storaged in the groundwater aquifer

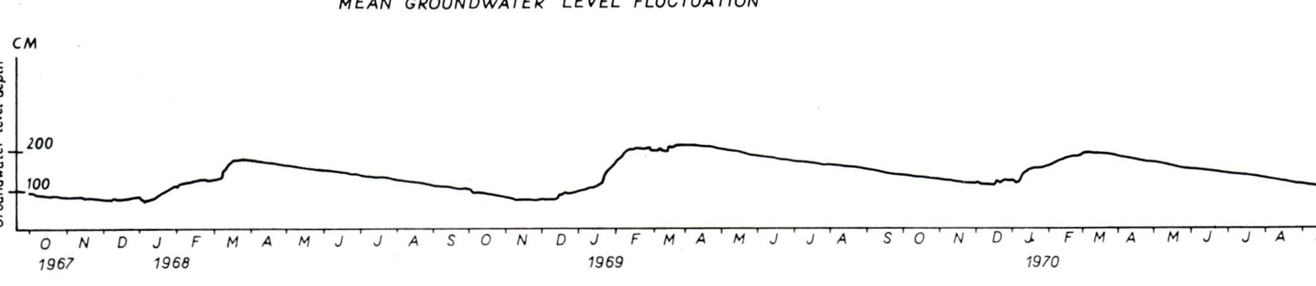

Fig. 5.4 Groundwater level fluctuation due to the evapotranspiration in the wet and dry region of Africa

for consumption during the dry period. The groundwater level fluctuation (Fig. 5.4) clearly confirms this, although originally it was thought that the decreasing part of the graph indicated the groundwater contribution to the baseflow.

5.3 GROUNDWATER ASSESSMENT

Knowledge of the groundwater regime under natural conditions is essential for its development. Groundwater survey, continous collection of data, additional measurements of springs, pumping tests, geophysical survey and the application of isotopes are costly and therefore for developing countries it has been recommended that all types of surveys should be designed with regard to the practical objectives.

The assessment of groundwater availability goes far beyond the stage of data collection. It involves the study of groundwater flow, an assessment of the hydrological balance and budgeting the aquifers. It also includes the forecast of future use of water resources based on various alternatives.

Kisiel and Duckstein found the following approach to be suitable for groundwater assessment on a regional and local level:

A) Regional level.

a) Identify the areas of natural recharge and discharge,

b) estimate quantitatively natural recharge and discharge,

c) determine the direction and velocity of flow,

d) estimate total storage,

e) evaluate the quality of stored water.

B) Local level.

a) Evaluate the geological structure from the viewpoint of drilling technology,

b) evaluate the effect of pumping on the surrounding areas,

c) evaluate the effects of various alternatives of well fields and of the time variation of the pumping.

Beside technical aspects economic and social aspects also have to be considered before the groundwater resources are exploited. Particularly the evaluation of future demands, should be based on the results of a recently-conducted comprehensive survey among water users. Such a survey is not easy in the tropics, bearing in mind the many difficulties arising from field work and the limitations of time and financial resources usually allocated for such a purpose.

In an example given by Dijon [10] such a type of survey was performed under the auspices of U.N.D.P. in the region of the Upper Volta where the main task was to carry out integrated inventories of villages, including data on existing wells, water use, to evaluate the adequacy of water supplies and of reasonable levels of water requirements for the future. Obviously criteria for such a survey cannot be uniform for a great number of countries and a sensitive approach respecting local customs is an important part of the survey. For instance, villagers in the region east of Victoria Falls refuse to use the water from wells and always prefer the water from the sandy beds of intermittent streams. Obviously any drilling of water for domestic utilisation would be highly ineffective there.

It appears typical of groundwater survey in the tropics that all work is directed towards the improvement of the rural supply. In wet and dry regions water supplies are often a decisive factor for the further stabilisation of village settlements. Therefore small groundwater schemes should be given priority in regions where an increasing shortage of water supply can be expected.

Zafiryadis [50] emphasized the need to provide small quantities of water by means of wells situated near the villages. The method of investigation based on experiences from Nigeria consists of two phases:

a) Individual investigation in each village,
b) extended hydrogeological investigation.

During the first phase it is essential to consider each village as a separate problem to which a solution must be found within a limited area around the village. This phase consists of the evaluation of existing information on areas around villages by conventional hydrogeological reconnaissance, interviewing the population and if necessary by aerial photography. At the end of the first phase either a solution can be recommended directly or alternatively the second phase has to be initiated. This usually consists of geophysical investigations and exploratory drilling. This is done with the objective of later converting the test wells into production wells.

In groundwater surveys there should be a different approach to humid and arid tropics. In a humid region flood control may have priority in water planning and this may lead to the conclusion that the groundwater is always in abundance or because of the availability of surface waters it is considered to be a less significant source. However, in these regions also groundwater often becomes the only source of supply during more or less prolonged dry period or when surface water becomes highly polluted and its treatment expensive. Groundwater however, is almost always a reasonable source of safe water supply for the villages and for domestic use. Groundwater is most significant in semi-arid and arid lands where it is often the only water resource available. Here even a costly survey on a large scale and including aerial photography, field geological survey, groundwater data collection and evaluation, geophysical investigation and istope studies, is usually profitable.

In efforts to find the critical amount which can be extracted from the aquifer as part of renewable water resources it is necessary to make use of existing observations. Only after they have been evaluated can newly established observational programmes together with the other survey methods be planned, mainly to fill the gaps in existing data and identification results. At present such a work is being directed towards the establishment of data banks. Haman [17] listed the aims which can be achieved through data banks as follows:

a) to provide a progressive build-up of knowledge of groundwater conditions in the area,

b) to provide a reliable background for water resources master planning and the selection of the most promising areas for groundwater resources development,

c) to optimize the number of wells, drilling depth and well construction,

d) to control groundwater quality,

e) to protect groundwater from pollution,

f) to improve the management of groundwater resources.

A practical example of such a utilisation came from the Metro Cabu area of the Philippines, where the main objectives were to establish reliable figures on the number of wells existing in the area and amount of withdrawal from different types of wells. These data with files of well records were stored in the databank as basic material which could be used for the management and control of the development of groundwater resources. Emphasis was placed on the legal aspects of the project. Permission for drilling had to be obtained before the drilling was started and thus efficient management of groundwater resources and control of withdrawal may be achieved by granting the right to withdraw a specified amount of groundwater from aquifers in the area. This is based on results of pumping tests with prognosis of water level decline as a function of time and distance for desired withdrawal. A significant part of the databank is the well record section established for the purpose of receiving, filing, and relasing data. A system identifying single wells and boreholes on maps, in the field and in computer language is another significant part of such a project.

The involvement of too many different organizations and agencies instead of a single one may become a serious barrier in groundwater development. In Thailand (Piancharoen [35]) groundwater is controlled by the Department of Mineral Resources, the Department of Public Works, the Department of Public Health, the Office of Accelerated Rural Development, the Metropolitan Water Works Authority and the Irrigation Department. Nevertheless, under limited control the piezometric head in the vicinity of Bangkok has decreased by 12 metres in the last thirteen years and with present exploitation a decrease of 100 metres by the year 2000 is forecast, while with possible protective measures it will be only about 60 metres.

Reports from other parts of the tropics also indicate that non-renewable water resources are being heavily consumed. In the groundwater assessment the differentiation between renewable and non-renewable resources is one of the most difficult tasks of modern hydrogeology. The problem must first be solved regionally, this being based on the complex water balance calculation taking into account the long-term mean of annual precipitation, evapotranspiration, surface runoff and the present volume of groundwater already exploited. In most cases, such as the question of the water balance of geologically complicated structures of considerable size, a model approach is highly recommended (Balek [4]).

The data for the simulated period should be selected carefully and accurately and extended by the use of long-term records of annual rainfall. The actual and simulated difference between the groundwater level at the beginning and at the end of the period selected for the simulation is a dominant indicator of the efficiency of the simulation (Fig. 5.5). Also a general agreement between the simulated groundwater outflow and actual total hydrograph is a good indicator of the model's reliability.

In contrast to other scientific results groundwater simulation routines developed in temperate regions can be applied in the tropics provided that the identification

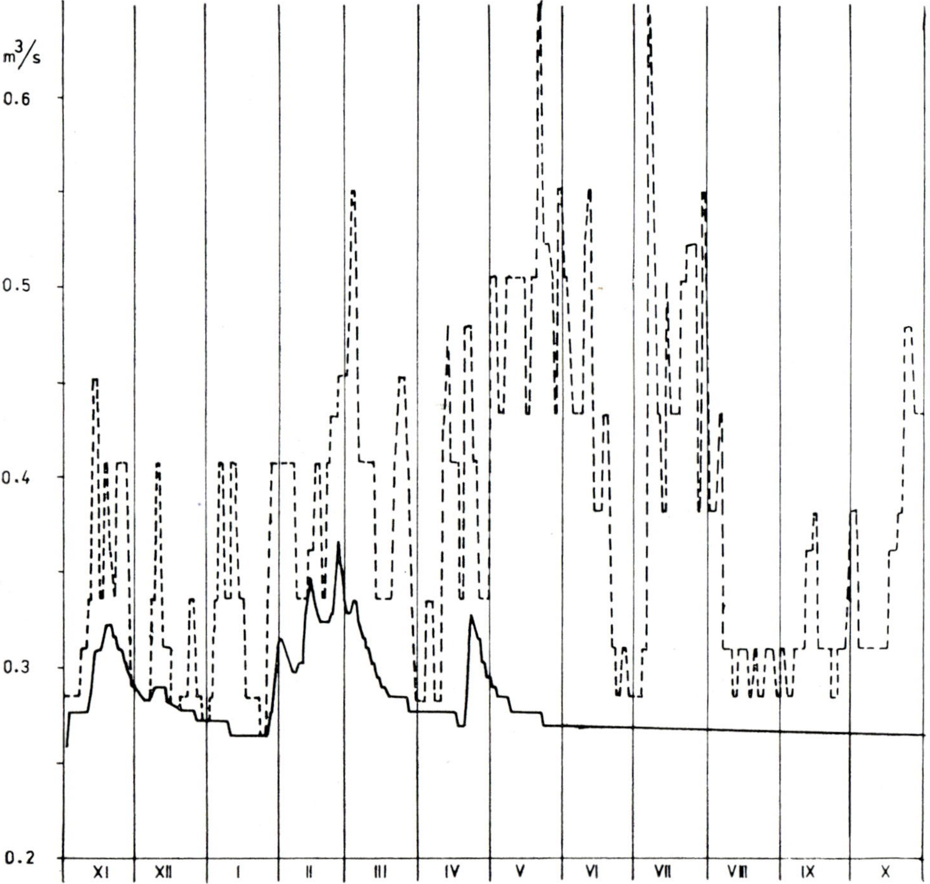

Fig. 5.5 Actual (full line) hydrograph and simulated groundwater outflow (dotted line) as used for the model optimization

of the system is based on data obtained from the simulated basin. A simplified flow chart of a possible approach is shown in Fig. 5.6.

The model work is directed toward the calculation of the amount of water which

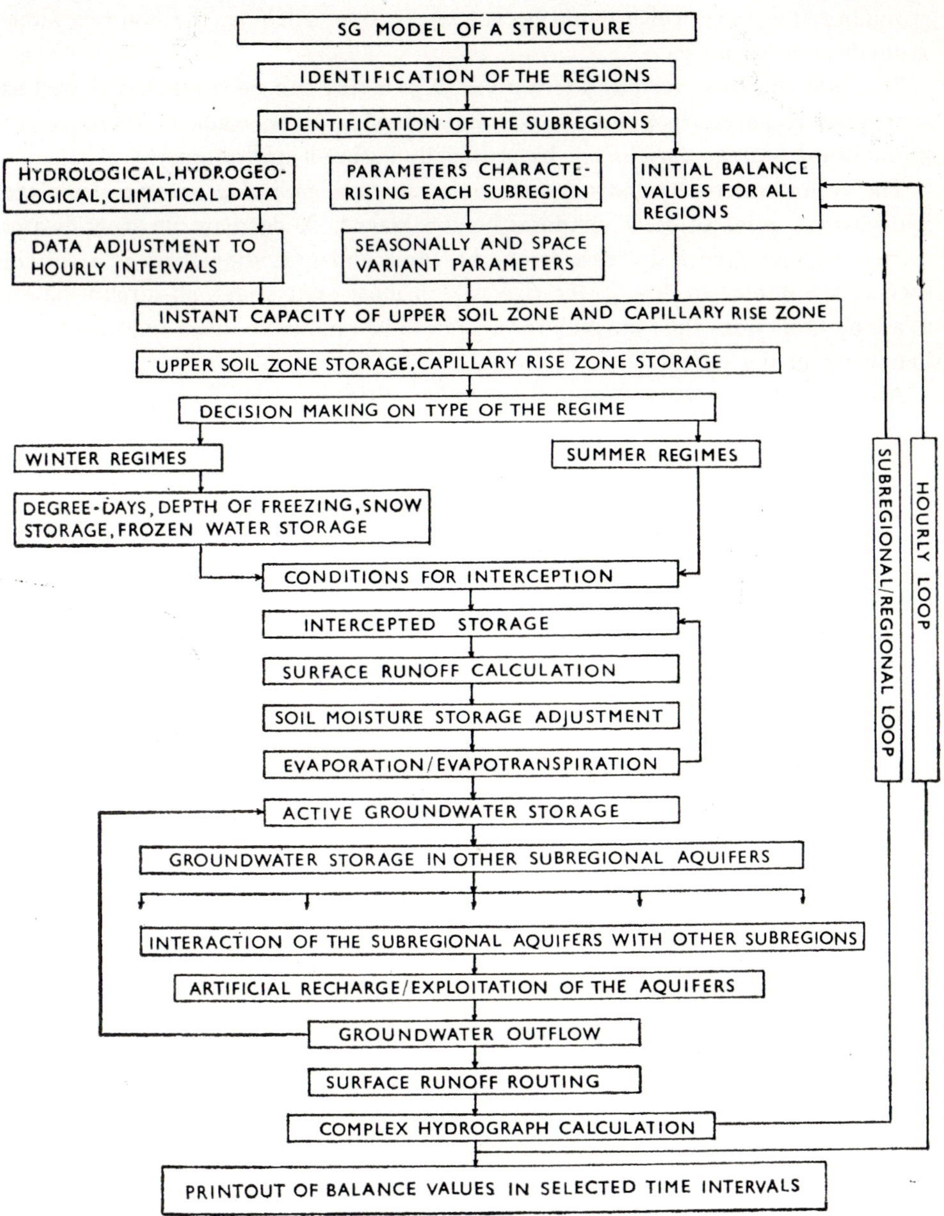

Fig. 5.6 Simplified flow chart of the water balance model for the groundwater recharge simulation

results in the formation of the annual baseflow and/or which is consumed from the groundwater resources by evapotranspiration. Providing that in a long-term mean the groundwater level is more or less stable or fluctuating within certain limits, both parts represent the groundwater recharge or in other words that part of the groundwater storage which could be utilized without affecting the non-renewable groundwater resources.

The final effects of various alternatives to pumping can be simulated as well as a negative recharge. In a similar way an analysis can be made of the effect of groundwater storage enrichment by artificially induced infiltration.

The groundwater – baseflow system under artificial pumping becomes from the hydrological point of view temporarily unbalanced. A decrease in groundwater storage will produce a decrease in the baseflow and providing there is a certain limit to the minimum flow in the river which must exist under all circumstances as a long-term mean the following formulae can be applied in order to estimate the behaviour of the system:

A simplified water balance equation can be written

$$P = E_t + R + O$$

where P is the long-term mean annual rainfall, E_t the mean annual evapotranspiration, R the surface runoff and O groundwater outflow (all values in mm). O also indicates the long-term groundwater annual recharge, providing there is no artificial recharge or depletion. If depletion of groundwater occurs a new value is introduced

$Z = O - M$ for depletion and

$Z = O + M$ for recharge.

Here M is also given in mm year^{-1}.

Under such an impact the water balance of the basin becomes disturbed. The groundwater outflow O' can be calculated as occurring after T hours of the impact

$$O' = \frac{1}{T}\left\{[8760G_i(1 - K) - Z]\frac{1 - K^T}{1 - K}\right\} + Z \qquad \text{mm year}^{-1}$$

where O' is the mean annual groundwater outflow volume (mm year^{-1}) after T hours, G_i is the initial groundwater storage, (mm) and K is the recession coefficient based on hourly intervals.

Mean groundwater outflow volume for the period of T hours is

$$\bar{O}' = K^{(N-1)}[8760G_i(1 - K) - Z] + Z \qquad \text{mm year}^{-1}$$

Finally it can be estimated the time N during which the groundwater outflow is stabilized at a fixed value of O_N mm year^{-1}.

$$N = \frac{\log\left[\frac{O_N - Z}{8760G_i(1-K) - Z}\right]}{\log K} + 1 \quad \text{hours}$$

These approximate formulas are valid for conditions of steady recharge and depletion and therefore some deviations can be expected for extreme dry or wet years.

As can be seen from the formulas G_i can be estimated for the conditions that $O = Z$ or in other words for the undisturbed natural conditions. Then

$$G_i = \frac{O}{8760(1-K)} \quad \text{mm}$$

The calculation of G_i and K is also discussed in Chap. 3. A natural recharge $Z = O$ of the undisturbed hydrologic system can be estimated to be a separated baseflow from the mean long-term annual hydrograph.

5.4 GROUNDWATER POLLUTION

In addition to the exhaustion of non-renewable water resources, the problem of groundwater pollution is steadily growing worse in the developing countries of the tropics. For instance Jacks [23] reports an increasing concentration of nitrogen in the groundwater of southern India to as much as 1500 mg l^{-1} for which the washing effects of monsoonal rainfall can be blamed. The industrialization of tropical developed countries resulted in groundwater pollution long ago. The countries in which the problem of groundwater pollution has only become serious recently, have the advantage of utilising the experience of developed countries to avoid the application of costly protective measures when it is too late.

A basic part of protective measures is the application of effective monitoring practices and their inclusion in a cost effective methodology. A proposal for a monitoring system for developing countries in the tropics has been established by Everett [12] as being suitable for all types of groundwater aquifers, areas and basins. In general, a typical monitoring program varies in accordance with the climate, hydrogeology, population and source of pollution. A typical program must focus on measurements relating to the most important sources of pollution.

The purpose of monitoring is seen in the assessment given to local agencies in the drawing up and implementation of protective programs and the establishment of priorities in protective measures, their location and financing.

The following are procedures for protective measures:

a) Selection of the regions to be monitored. Here administrative, physiographic and priority aspects play a role. Political aspects may arise where the boundaries cross major river basins and pollutants from the headwaters cannot be adequately monitored. Usually areas with the largest sources should be selected first as is the case of the River Kafue in Zambia, flowing from heavily industrialized regions of the mining industry.

b) Identification of sources of pollution and their causes as well as methods of waste disposal. Here the identification of underground causes in particular may prove to be difficult.

c) Identification of the most frequent potential pollutants:

- physical: temperature, density, odour, turbidity;
- inorganic: chemical, major constituents, other constituents, trace elements, gases;
- bacteriological: coliforms, pathogenic micro organisms, enteric viruses;
- organic: carbons, organic matter, methylene blue active substances, nitrogen compounds, chemical oxygen demand, phenolic compounds, pesticides, herbicides.
- radiological: alpha activity, beta activity, strontium, radium, tritium.

d) Identification of groundwater usage. There should be an evaluation of the quantity of groundwater being extracted, the location of major pumping centres, the volume of polluted waters and their association with streams, the impact of irrigation, deposits of waste into dry stream beds and the volumes associated with water infiltration into the soil from ponds, seepage pits, tanks and from various sources of agricultural pollution.

e) Definition of various hydrological situations which may develop or accelerate pollution processes.

f) The study of existing water quality. The quality of the water before pollution took place should always be identified. It is clear that a successful experience in one country cannot always be repeated elsewhere, where there may be a different environmental or population situation. In many parts of the tropics the quality of present water resources is far above the quality of drinking water as given by the standards of the industrialised countries.

g) Evaluation of the infiltration potential of waste on the land surface. Here it is essential to analyse the hydrogeological system from the point of view of water passing through the vadose zone into the zone of saturation. This may be found to be one of the most difficult points and many simulating methods and models have been developed with varying success to identify the whole process.

h) Evaluation of the mobility of pollutants from the land surface to the water table.

i) Evaluation of the attenuation of pollutants in the saturated zone.

j) The establishment of an inventory of pollution sources and causes.

k) Evaluation of existing monitoring programmes if any. The main aim of this is to evaluate all deficiencies in the programmes. The final result should be a list of important sources and causes for which a monitoring system should be established.

l) The establishment of alternative monitoring approaches. All possible factors including cost should be taken into account.

m) The selection of monitoring programmes should cover priority causes and sources, monitoring gaps and verifying existing programmes.

n) Criteria for reviewing and interpreting the results of monitoring.

o) Summary and transmission of monitoring information. Usually priority is given to federal, state and local agencies. Information to the public is regarded.

A serious problem in regions with long coastal line and shallow aquifers is the intrusion of marine water into the exploited wells. Cuba is cited as an example where intrusion of marine water has an impact of from 2 to 15 kilometres inland. In such a region, the safe yield should be analysed and calculated not only with respect to the quantity of water, but also its quality. In other words, accurate calculation of the maximum groundwater table depression has to be provided with regard to the elimination of the negative effect of marine water intrusion. Shayakubov and Morales [39] proposed the formula:

$$S = h_n + T - \frac{M + H}{\alpha}$$

where $\alpha = \dfrac{h_c}{h_n}$.

In the formula S stands for the admissible groundwater table depression, H is the depth of the well, T is the depth of the well to the sea level, M is the safe distance of the intrusion source from the bottom of the well, h_c is the depth of the intruding marine water below sea level, h_n is the head of the groundwater level above sea level.

Many potentially usable groundwater resources are subject to additional chemical constraints. In karst regions especially, this can become a serious problem. For instance in Yucatan, Back and Lesser found that water-bearing aquifers are in the limestone formations and the groundwater is concentrated in caves and cenotes. A thin layer of freshwater located only few centimetres above sea level is the only source of water for several thousands of inhabitants. The chemical composition of the water is determined by the mixing of rain with subsurface salt water, solutions of limestone and gypsum and contamination by organic material and sewage. Limiting factors resulting in water scarcity are extensive karstification which with rapid infiltration prohibit the formation of surface sources, the uneven seasonal distribution of rainfall and long-term permanent pollution from organic and inorganic sources.

Such an example only illustrates the complexity of groundwater problems in various parts of the tropics and any solution if found, cannot be directly applied outside the region studied.

A complex study of the chemical composition of groundwater and seepage was carried out in west Africa by Roose and Lelong [38]. Based on data analysis from eight stations representing various types of natural soil – vegetation ecosystems, the following conclusions were reached for the ITCZ zone:

a) The influence of the bioclimatic differentiation on mean chemical composition is only slight,

b) the amount of dissolved chemical species is highly variable according to the seasonal pattern of rainfall and flow volumes,

c) the chemical composition of water in the soil layers is controlled by biological and biochemical processes, while the phreatic water is controlled by physico-chemical conditions,

d) mineralisation is increasing from rainfall and throughfall waters to runoff and to drainage water, but there are decreases in spring waters, except for Si and Na.

5.5 GROUNDWATER IN THE TROPICS

An attempt to evaluate the groundwater resources of the continents was made in Soviet laboratories [47]. Groundwater resources are recognized separately in three zones:

a) an active groundwater zone in the upper part of the earth's core to a depth of 200 metres and above the erosional base.

b) a less active zone, however, with a depth above sea level, interacting with various extensive depressions and deep valleys.

c) zone below the surface of the oceans to a depth of about 2000 metres below sea level.

The groundwater storage was estimated for entire continents and separate values for the tropics are not available. As can be seen in Tab. 5.1, the richest resources of the first type are found in Asia, followed by Africa, Latin America and Australia/

Tab. 5.1 Groundwater resources of the continents as estimated by Soviet water resources research laboratories

Continent	Groundwater zone (no.)	Depth (m)	Effective porosity (%)	Storage (10^6 km^3)	Total
Asia*)	1.	200	15	1.3	
	2.	400	12	2.1	
	3.	2000	5	4.4	7.8
Africa	1.	200	15	1.0	
	2.	400	12	1.5	
	3.	2000	5	3.0	5.5
South America	1.	100	15	0.3	
	2.	400	12	0.9	
	3.	2000	5	1.8	3.0
Australia/ Oceania	1.	100	15	0.1	
	2.	200	12	0.2	
	3.	2000	5	0.9	1.2

*) Including temperate and polar regions.

Oceania. The groundwater resources of the second and third types are of the same order.

Total baseflow from the continents is shown in Tab. 5.2. Here the situation is different. The greatest proportion of world baseflow comes from Latin America, followed by Asia and Africa. In fact only a small part of the groundwater zone contributes to the formation of the baseflow.

These estimates have to be regarded as being only approximate and the present complex survey of tropical groundwater resources is far form being completed. A more complex and systematic evaluation has been accomplished in Australia. Sedimentary basins underlie about 60% of the Australian continent. The aquifers extend over large areas and are of considerable thickness. As stated by Hughes [22] these aquifers have proved to be of great value in the development of the Australian economy by providing the sole significant source of water in extensive arid and

Tab. 5.2 Groundwater outflow from the continents as estimated by Soviet water resources research laboratories

Continent	Groundwater outflow (km^3 $year^{-1}$)	Percentage of total outflow (%)
Asia*)	3750	26
Africa	1600	35
South America	4120	35
Australia, Oceania	575	24

*) Including temperate and polar regions.

semi-arid areas. Tab. 5.3 lists most of Australia's significant artesian basins and Fig. 5.7 shows their location.

The Great Artesian Basin is considered to be the greatest artesian basin of the world and it is of great importance particularly for Australia's pastoral industry. Total withdrawal from it is estimated to be 540×10^6 m^3 $year^{-1}$ or 17 m^3 s^{-1}. The other aquifers supply a total $500-1000 \times 10^6$ m^3 $year^{-1}$ or $16-32$ m^3 s^{-1}.

The Murray basin has several outlets with a total estimated discharge of about 3 m^3 s^{-1} and drains partly into the sea, shallow unconsolidated aquifers in alluvial deposits of ancient river valleys, coastal dunes and shoreline deposits are also important as are fractured rock aquifers. Except for the western parts of Tasmania and some regions with perennial streams an increase in groundwater consumption to meet rising irrigation demands has been observed. The total amount of artesian

Tab. 5.3 Main Australian artesian basins

Basin	Area (10^3 km^2)	Depth (m)	Mineralization (g l^{-1})
Great Artesian	1751	2100	6.2
Murray	282	30—400	1.7
Canning	388	30—550	0.3
Eucla	191	90—600	6—35
Northwestern	78	60—1200	4—5
Coastal Plains	54	60—750	
Gulf	31	60—300	

boreholes and wells in Australia is estimated to be 200 000 with their depths reaching 2000 metres.

Great attention is paid to dry river beds where groundwater can be utilised for stock. In New South Wales attention is paid to the groundwater in flat regions flooded during periods of rainfall.

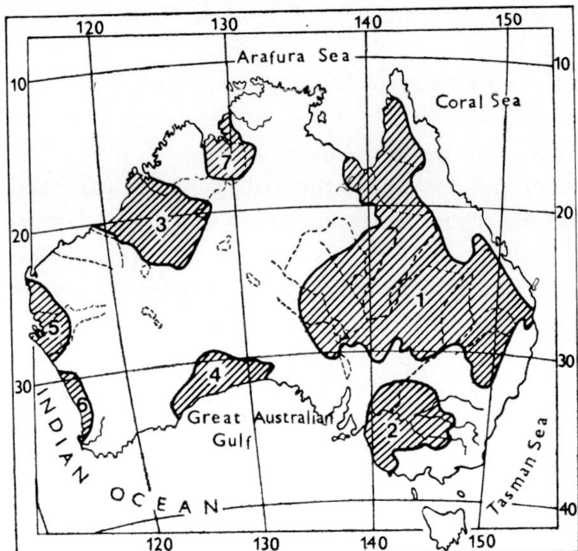

Fig. 5.7 Main Australian groundwater basins. 1, Great Artesian; 2, Murray; 3, Canning; 4, Eucla; 5, North-western; 6, Coastal Plains; 7, Gulf

In western and southern Australia, in the absence of better groundwater resources cavities in granitic formations known as gnamma are explored.

Mound springs are important in some localities but because of their difficult accessibility, methods of aerial photography were applied to trace them (Williams and Holmes [46]). By correlating the discharge with the surrounding swampy vegetation, supported by the springs, it was found that the spring water is dissipated by a mean evaporative flux of 5.5 mm day^{-1}. The term mound is related to a distinct pattern caused by the accretion of sediments around the outlets. The height of the mounds occasionally reaches more than 10 metres.

In the Great Artesian Basin some groundwater samples were found to be as much as 350 000 years old. The basin consists of a multi-layered confined system with aquifers in continental deposits of quartzose and sandstone from the Triassic, Jurassic and Cretaceous periods. Recharge occurs mainly in the eastern zone and natural discharge occurs from springs which are connected to structural features and from mounds. Airey et al. [1] concluded that variations in the rate of infiltration into the aquifers correspond to the glacial and interglacial periods.

The most serious water quality problems in Australia, which at some time may appear elsewhere, were found to be:

a) salinization of soil in the major irrigation areas of the Murray valley resulting from long-term irrigation and shallow water table;

b) high nitrate concentration in the limestone aquifers of southeastern Australia as a result of waste disposal from dairy factories;

c) migration of faceal bacteria and phosphates from septic tanks into shallow sand aquifers in the Perth region;

d) pollution of the multiple basalt system west of Melbourne by the organic waste;

e) potential contamination associated with agricultural practices in Victoria;

f) intrusion of seawater into coastal aquifers,

g) disposal of iron-rich acid effluents in western Australia.

Very special groundwater problems are relevant to the hydrogeology of tropical islands. Taylor [40] studied these problems in Pacific region and found typical features to be a lack of natural surface water in the form of freshwater lakes, ponds and permanent streams. Sand and limestone and the steep orography of the islands ensure the rapid movement of rainwater. Most of the water is consumed by abundant vegetation. The annual rainfall total is highly variable being between 166 mm for Onotoa island and 6732 mm for Funafuti. Dry spells continue for about three years followed by wet spells. Both may last for several years however, which makes conditions very much less favourable for the formation of significant groundwater resources.

The greater part of the African continent consists of a Precambrian crystalline and metamorphic shield with three main units. The outcroppings of the basement correspond to the uplift areas which encircle the depressed basins with sedimentary formations such as those of the Niger, Chad, Nile, Congo and Kalahari. Cambrian and Infracambrian formations of marine origin, composed mainly of calcareo-dolomitic rocks and calcareous shale have been identified at lower levels and are found mainly south of the equator at Katanga and metamorphosed in Zambia. The younger formations consist of shale and sandstone with some limestone and clay beds. The Somalian plateau and some coastal sedimentary basins were once covered by the Jurassic and Early Cretaceous seas. During this period called the Continental Interclaire, formations of sandstone were deposited in most of the basins. During the Tertiary period, formations of sandstone were deposited. During the Miocene and Pliocene periods eruptions occurred on the breaking lines along the Rift valleys.

Precambrian formations of large peneplains at low altitudes in west Africa and at higher altitudes of the east African highlands have their water-bearing units in weathered fractures of combined sections. The upper horizon is frequently formed by the laterites and argilosands temporarily containing groundwater. The next horizon is formed by kaolinic porridge. Below are weathered sections of bedrock holding water under a certain head, so that the effectiveness of the wells is related to the locality chosen. In contrast to schists, the quartzites, dolerites and vein-rocks have better yields.

Important water bearing units are found in Infracambrian and Paleozoic formations. Single boreholes producing 100 m^3 $hour^{-1}$ are reported.

The large sedimentary basins of Central Africa are filled with soft Paleozoic, Jurassic and Cretaceous deposits and some of them, as in the Chad region and northern Nigeria, contain strong resources of artesian water.

Sedimentary basins of the west and east coast and parts of Malagasy occasionally have a high yield, however, near the coast fresh water may be mixed with seawater.

The water yield of the Mesozoic plateau of Ethiopia and Somalia is highly variable.

The Resources and Transport Division of the United Nations prepared a guide (Tab. 5.4) giving basic information on the water yield in some geological formations of Africa.

Tab. 5.4 Estimated yields of selected African aquifers

Formation	Average yield (m^3 $hour^{-1}$)	Maximum yield (m^3 $hour^{-1}$)
Precambrian		
Granitogneiss	2—5	20 and more
schists, rhyolites, diorites	less than I	no information
Infracambrian and Paleozoic		
Shale-limestone and calcareo dolomite	10—100	no information
West African shales and shale-limestones	no information	no information
Large sedimentary basins of Central Africa		
Karoo sandstones	1	200
Artesian waters of northern Nigeria	no information	1000
Kalahari sands	1—10	50
Congo and Chad basins	1	no information
Continental Interclaire	no information	100 and more
Alluvium	small	up to 1000 in Nile region
Sedimentary coastal basins	20	300

A great part of the groundwater resources in Africa are of ancient origin. For example, the results of ^{14}C dating in Zambia confirmed the age of water in lower aquifers as being about 4000 years old (Burgman et al. [8]).

Geygh and Wirth [14] found the age of water in northern Nigerian Cretaceous and Tertiary formations to be from 3000 to 30 000 years. The fossil water in so-called Gwandu aquifers was found to move 150 metres per year.

In addition to special surveys, more typical ones have been the groundwater studies which form a part of integrated hydrological studies. For example, during a complex hydrological survey of great African lakes basins, performed under the patronage of the W.M.O. one task was to estimate the total groundwater storage. McCann [31] suggested the application of a simple formula for the region to estimate the volume of groundwater in the fissure system:

$$V = PAtB$$

where V is the volume of groundwater in the fissures (m^3), A is the drainage area of the basins (km^2), t is the thickness of the fissure system (m), P is the porosity of the rock units containing the groundwater and B is the percentage of the basin underlined by the fissure system. Taking both latter values by 0.01 and 0.7 respectively, the groundwater storage was estimated as shown in Tab. 5.5.

Tab. 5.5 Groundwater storage in the basins of Great African lakes, according to McCann [31]

Basin	Drainage area (km^2)	Storage (10^9 m^3)
Lake Victoria	202 300	70
Lake Kyoga	69 500	24
Lake Mobutu	18 700	7
Total	290 500	91

It was concluded that only a small portion can be withdrawn for use. As stated by the author, for visualizing the usable amount, it represents 50% of total precipitation in a year.

In Saudi Arabia an area of 800 000 km^2 was surveyed by using $^{18}0$, deuterium, tritium, ^{34}S, ^{13}C and ^{14}C. The analysis of 400 samples indicated that a considerable part of present water resources was recharged about 20 000 to 25 000 years ago, while very little is currently being recharged. About 20 mm of present downward movement was traced by tritium, however, it was not clear whether the water actually reached the upper aquifer. The complexity of possible interaction between deep and shallow aquifers which was studied at Quatar (Yurtsever, Payne [49]) makes detailed studies very difficult. Again in the coastal regions the intrusion of seawater makes the system even more complicated.

In Chile the age of the groundwater was determined to be 5000 to 11 000 years at various locations. In Pampa del Tamarugal the age of the groundwater was between 16 500 and 33 000 years (Fritz et. al. [13]). In this near-desert area a minor subsurface recharge occurs at the foot of the Andes and possible source today are the exogenous zones of the Andes. Water comes from surface infiltration of flood water through creek channels (Fig. 5.8).

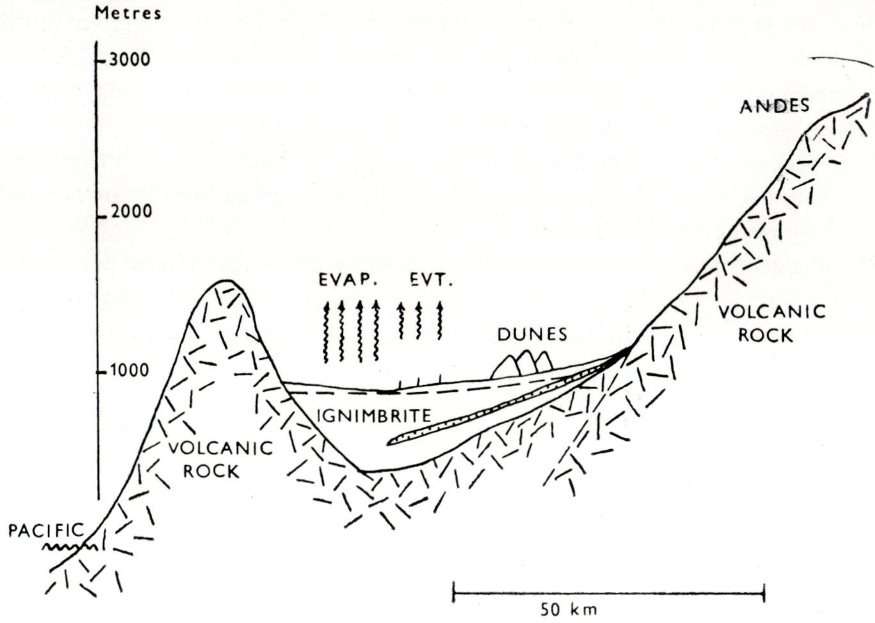

Fig. 5.8 Groundwater recharge in Pampa del Tamarugal, according to Fritz et al.

In this region of South America another significant source of water was found in the regions where shallow alluvial layers transverse valleys descending from the Andes. Lloyd [27] assumed that under such conditions the prevailing steep groundwater level gradients mean that groundwater resources are closely related to seasonal recharge. Nevertheless, at present the areal aridity is steadily increasing and water supply problems are becoming more serious.

In Mexico the situation is in many ways typical of that in other Central American regions. Groundwater, especially in semi-arid and arid regions, is the most important water resource. The largest consumption of water is by irrigation systems. In fact 50% of national agricultural production depends on groundwater resources. As estimated by Gonzales [15] more than 30 significant aquifers in Mexico are over-exploited at present.

In Venezuela significant water-bearing aquifers are found near Maracaibo, however, the total size is only 1300 km^2 and approximately 35 km^3 of water, aged from 4000 to 35 000 years, are storaged there.

Many springs are found at the foot of the Andes and it is assumed that they have recharge areas at high altitudes and that water is transported mainly through fractures in tuffs and lava flows. Isotope analysis indicates that many of the present springs are recharged by water stored under slightly but still significantly different climatic conditions. At some locations where the potential recharge areas have very little recharge today it is expected that the regime of springs may represent diminishing hydrologic systems which no longer receive any significant recharge. A similar conclusion has been reached for the baseflow of some rivers in the region, especially for those which show displacement from the meteoric water line.

It is believed that a significant amount of groundwater in the arid regions of Latin America is lost due to evaporation, as the groundwater in the lower reaches of the valleys is close to the surface.

In India also, groundwater is of primary importance for agricultural development and rural water supply. As stated by Baweja [6], about 80% of the Indian population are dependent on agriculture, therefore 90% of total usage is accounted for by irrigation. Of 572 000 villages more than 50 % do not have a safe water supply. In 1974 119 billion m^3 of groundwater were utilised. In 1980 total usage has been estimated to be 275 billion m^3. The main advance in groundwater exploitation has come about in the alluvial areas where the construction of wells and boreholes is relatively simple. However, 70 % of India consists of hard rock formations where the groundwater is tapped by wells that have been dug.

Spontaneous consumption, not based on knowledge of actual recharge, resulted in a sharp fall of the groundwater level. Therefore, hard rock formations are given priority by Indian groundwater surveyors. Trap basalt formations at Maharastra, Andhra Pradesh crystallines and Karnataka crystallines are promising. Attention is paid to alluvial formations in Uttar Pradesh with regard to the possibility of enriching present groundwater storage through increased seepage from canals and through the infiltration of excess monsoon rainfall (Balek [5]).

A rapid decrease of groundwater storage is reported from various parts of India. In the regions Tamil Nadu and Kerala, formed by granites, charnockites, gneisses and schists, the wells have been deepened because of a continuous decrease in the groundwater level by some 10 metres.

Athavale et al. [2] used tritium injection for the estimation of recharge to phreatic aquifers in India with the aim of determining the percentage of total rainfall contributing groundwater recharge. They found that only about 8% of monsoon rainfall (some 1250 mm) was recharged into the aquifer. The actual value varied between 50 and 240 mm this depending on the percentage of sand in the soil (Fig. 5.9). The development of the post-monsoonal evapotranspiration process can significantly influence the final recharge effect.

Geophysical prospecting has been found more effective in many parts of India, especially in the basaltic rocks where weathered zones are scarce and water is limited to fractures and fissures. As has been shown by Bose and Ramakrishna [7], geophysial prospecting can be recommended particularly in the formations of the Deccan traps where residual soil and latterites are capping over the traps, in weathered zones, in fissure/fracture zones, in vesicular trap horizons, in inter-trappean beds and in infratrappean formations.

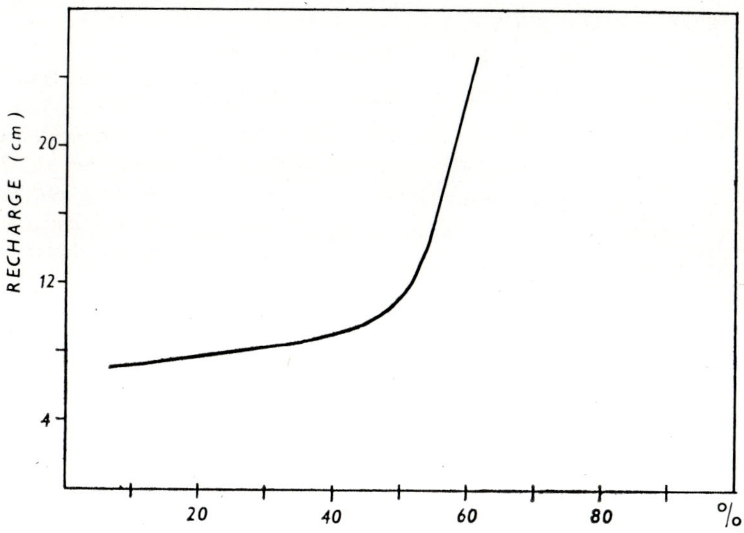

Fig. 5.9 Relationship between average sand content in soil and annual recharge in India, according to Athavale

5.6 GROUNDWATER IN DESERTS

According to Dixey [11] the aridity and humidity of an area depends not only upon the climatic conditions of the area itself but also upon climatic conditions in the surrounding regions and upon geological and geomorphological conditions. When these are of a late geological age it is clear that they can be of considerable importance in the present context with the existence of a prevailing desert regime. It is quite possible for geological and geomorphological processes to alter profoundly the climate of parts of a region. This would be reflected in a lack of harmony between the observed features of the soils, water supply and the current climate. For instance, geomorphological uplifts in many parts of Africa may have resulted in increased precipitation and a decrease in aridity. Lake Rudolph in a very arid region, was created by a gentle downwarping.

Artesian waters have a special importance in the arid tropics, though their utilisation has not yet been fully developed. African areas with a precipitation between 0 and 250 mm have one of the world's greatest aquifers. As estimated by

Nace [32] in the Nubian sandstones and Continental Intercalary Fountain, about 600×10^3 km of water are stored in an area of 6.5×10^6 km^2. The current recharge of these reservoirs is negligible and they were formed during the pluvial period at the end of Pleistocene, which means that the age of the water is 30 000 to 40 000 years (Fig. 5.10). Isotope studies have proved that the groundwater is more than 20 000 years old. The interruption of the recharge, as indicated by isotopes, occurred 16 000 – 20 000 years ago.

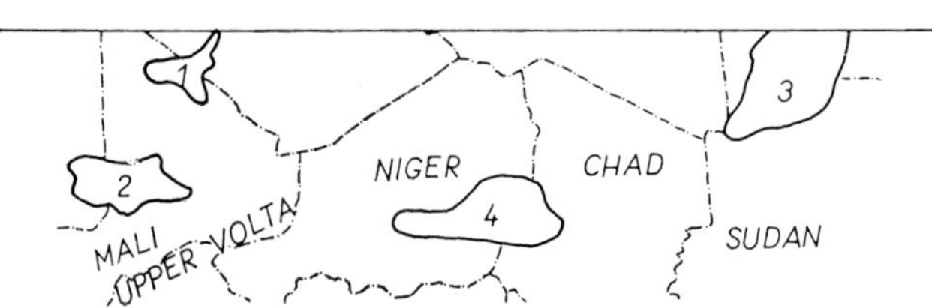

Fig. 5.10 Distribution of groundwater in tropical Sahara. Storage capacities estimated for: 1, Tanezrouft 0.4×10^3 km^3; 2, Niger 1.8×10^3 km^3; 3, Western Egyptian desert 6×10^3 km^3; 4, Chad 3.5×10^3 km^3

Either natural or drilled artesian water outflows, forming lakes or swamps, are known as shoots in Algeria and shebkas in Libya. The sources have been occasionally tapped by hand by the local inhabitants down to depths of almost 150 metres. In the Sahara oasis water comes from two main groundwater horizons, namely from the Continental Intercalation and from the Continental Terminal. The permeable strata in the first horizon collected a great deal of water more than a hundred millions years ago. The seas covered this reservoir in cenomanian times and deposited an impermeable layer of clay and dolomitic limestone. The sea withdrew at the beginning of the Tertiary period about 70 million years ago and new continental sediments accumulated and formed the second horizon with equally considerable reserves of water. The land level varied after the end of flooding cycles that have affected north Africa since the Lower Miocene period. The northern Sahara was divided into two parts by the Mzab ridge. The first horizon is not far below the surface to the west, so that the water is not artesian while to the east it is very deep and the water is under a great pressure.

It is estimated that by utilising 10 m^3 s^{-1} of the Sahara's fossil waters, no more than one ten-thousandth of the resources should be consumed in the next twenty years.

From another point of view, the deep artesian wells of the lower Sahara aquifer which penetrate to a depth of 1700 – 2000 metres are of economic interest if they can produce 400 – 500 l s^{-1}.

In the region of Chad the horizon is from 90 – 400 metres below the surface in an area of 100 000 km^2.

In the Kalahari Desert as in other semi-arid and arid regions it is expected that practically all groundwater located there is of fossil origin and no significant recharge can occur at present. Mazor et al. [30] say that there are three significant aquifers in the region, namely a phreatic aquifer in the sandy beds, a heterogeneous aquifer in the underlying and highly-jointed basalt and a confined aquifer in the deeper lying Cave sandstone. At least the lowest of these, which probably only has a small exchange of water with the upper ones, may provide a site for active recharge after the intensive withdrawal of water. The authors believe that recharge will occur by direct rain infiltration and that the artesian pressure in the Cave sandstone is caused by the phreatic water head situated in the Kalahari beds (Fig. 5.11). A relationship between the infiltration and annual rainfall has been

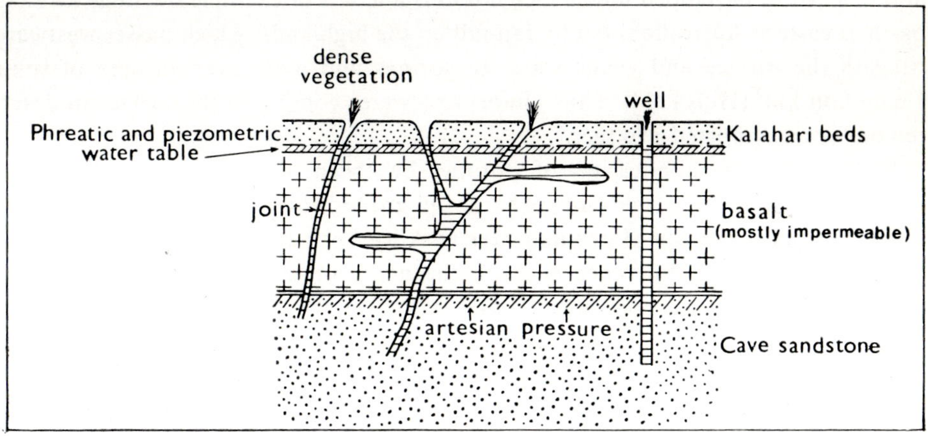

Fig. 5.11 Schematic representation of proposed groundwater regime at Orapa, according to Mazor et al.

observed in the Kalahari by Verhagen et al. [43], however they considered the recharge into the sand to be equivalent to the recharge into the aquifer.

In the western Kalahari the groundwater denitrification process has also been studied, where Vogel et al. [45] have attempted to ascertain the rate of groundwater denitrification. Remarkably, this region although sparsely populated and having no industry and relatively low cultivation, has a generally high concentration of nitrates, which under such conditions are of natural origin. The authors assume that the nitrates come from the atmosphere and are concentrated by evaporation, or formed in the soil by bacterial nitrogen-fixation. Common trees in the region such as *Acacia erioloba,* together with grasses, may be fixing plants. The groundwater age is estimated to be up to 27 000 years and it has been estimated that the initial nitrate content has remained between $1.6-0.7$ meq l^{-1} during the entire period. It has been concluded that the denitrification of such a type of aquifer takes 14 000 years to reduce all the nitrate.

Heaton and Vogel [18] used what is known as 'excess air analysis' to determine the groundwater recharge process in South Africa. The method is based on the indication that dissolved gaseous nitrogen and argon in groundwater is higher than the amount for the atmospheric equilibrium. The authors found that this phenomenon is widespread in South African aquifers and that it may indicate some intermittent recharge caused by the fluctuation of prolonged dry periods and periods of heavy rainfall, because variations in the amount of dissolved excess air are related to the physical structure of the capillary zones.

In many desert areas the saturated zone is frequently formed in more extensive areas than individual topographical basins. Both confined and unconfined types of aquifers can be developed under such circumstances. For example the huge artesian basin of eastern Australia is fed by rainfall on the highlands, which passes westward beneath the surface and yields water to springs and wells over an area of some 1.6 million km^2 (Hills [21]). Thus almost any water coming to the surface in desert regions is an imported product.

In Australia the application of isotopes has been widely distributed because almost everywhere the age and recharge of groundwater is of vital importance. Calf [9] studied the age of groundwater at Alice Springs, Northern Territory. He found no significant difference between the depth of the boreholes and the percentage of modern carbon, which indicated that the recharge had occurred at intervals of a few millennia, according to the variability of the climatic conditions. As in many semi-arid and arid regions elsewhere in the tropics, the process of infiltration obviously occurred at discrete time intervals separated by long periods of time when evapotranspiration vastly exceeded rainfall. Similar results were obtained independently by Pearson and Swarenski [34] in Kenya and by other authors already cited. Separate events can be proved by plotting ^{14}C data against the reciprocal of the total carbonate content (Fig. 5.12) each separate line indicating a single period of recharge.

By combining such results from both northern and southern parts of the tropics it can be concluded that evidence exists that periods of increased rainfall and arid conditions are very typical for all the tropics. However, still more extensive research is needed to find out whether these periods occurred at the same time in the whole belt.

There are several means of increasing water availability in the semi-arid and arid regions. Dijon [10] listed the following:

a) reduction of direct evapotranspiration by deepening the groundwater table,

b) change in cultivation which properly applied may reduce transpiration from the saturated zone,

c) development of devices facilitating the treatment of polluted water otherwise not usable,

d) artificial recharge,

e) optimization of water extraction, particularly through the proper selection of either open hand-dug wells or drilled boreholes. Although construction of the former is slow, they do not need specialized technical services such as the pumps required for drilled boreholes.

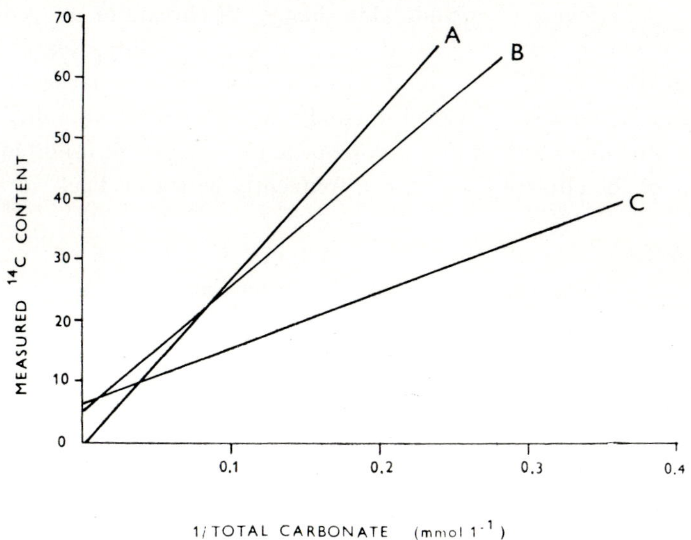

Fig. 5.12 ^{14}C measured content versus reciprocal total dissolved carbonate content, as an indication of different periods of recharge, according to Calf

Also the measures to protect groundwater resources as planned in Mexico (Gonzales [15]) can serve as a guide:

a) keeping permanent records of groundwater utilisation,

b) definition of restricted zones for controlling the withdrawal and exploitation of groundwater,

c) establishment of a centralized administrative unit responsible for all protective measures,

d) construction of artificial recharge installations whenever economically feasible,

e) protection of groundwater quality,

f) combined utilisation of surface water and groundwater whenever possible.

g) strengthening of the groundwater scientific survey,

h) cooperation among organizations in charge of various aspects of groundwater management,

i) reasonable utilisation of groundwater resources.

5.7 WATER IN SOIL

The vertical movement of soil water provides an important interaction between groundwater and vegetation or, in the absence of vegetation, as in the arid tropics, between the groundwater and the atmosphere. Climate is widely accepted as being an important factor in soil formation, nevertheless, several factors play an equally significant role. Because of considerable changes of climate in the geological past, types of soil which once originated under tropical conditions are described as tropical soils though they are now located outside the boundaries of the tropics. This category includes soils formed in conditions of extreme humidity and aridity. Conversely soils more typical of a non-tropical climate can be found in the tropics. The impact of the climate on soils can frequently be traced back to the Tertiary era.

As an initial source of information on types of soil when no better data are available, Lang's factor has been used. It is defined as

$$L = \frac{S}{t}$$

where S means mean annual precipitation (mm), and t is mean annual temperature, °C.

$L > 160$ is typical for the process of podzolization,
160 – 100 for the chernozem process,
100 – 60 for brown and chestnut soil formation,
60 – 40 for laterites,
< 40 for desert soils.

In modern terms more attention is paid to the meso-and microclimates, particularly as related to the dominant vegetation cover. The temperature and humidity regime is studied together with the wind regime. The last factor is particularly significant in arid parts of the tropics. Orography is another factor playing a significant role in the long-term process of soil formation. Finally, human activity is a factor of steadily increasing importance, unfortunately this factor frequently has a negative role. Fertilizing, overgrazing, uncontrolled cultivation and pollution of the soil rapidly alter its quality.

Climatic factors have direct impact on:

a) the physical properties of soil,
b) chemical weathering,
c) biological weathering,
d) physical weathering,
e) soil erosion.

Biological activity under extreme conditions may affect the soil processes very rapidly, especially when the rate of mineralization exceeds the production of organic matter, as in the humid tropics.

The weathering of alkaline and neutral soils above the groundwater level and near to it, results in the transport of silicic acid and the transport of CaO, MgO, K_2O and Na_2O while Al_2O_3 remains in a crystallic form as a residue, together with limonite, fragments of feldspars and other resistent minerals. In this way laterites are formed. The primary process of laterization can be transferred to the process of the resilication and the residue becomes enriched with SiO_2.

The process of laterization is not typical on acid rocks. Instead caolinitic clays are formed, often jointly with quartz. The process can be combined with desilication and with the transport of SiO_2, which results in the formation of concretions and bauxite surface layers.

Obviously the weathering process on the same type of parent rock can occur in different ways and the water regime of groundwater and soil moisture are the most decisive factors. For instance, in regions with high annual rainfall and high percolation rates, such as Indonesia, the products of weathering are intensively washed through the profiles and the clayish particles cannot become part of the soil profile. Thus on one hand the soil water regime affects the formation of the soil, and on the other becomes affected by changes in the soil profile.

The weathering processes in arid regions are often limited to physical weathering. When desert soils have not originated on loamy sediments, the content of clayish particles is very low. Under the additional effect of a certain type of wind, granularity of a certain type can prevail in the soil structure.

The process of humus formation is of a very special kind in the tropics. Despite the relatively rich vegetational cover in many parts of the tropics the humus content of tropical soils is not adequate. Here the orographic effect is different from that in temperate regions, because with rising altitude the humus content increases up to a certain level and from a certain altitude, varying according to latitude, it decreases again. In lower parts of the tropics the process of mineralization is much faster and is associated with temperature. It was estimated by Kutílek [26] that at a mean annual temperature of 24 – 30 °C the mineralization rate becomes higher than the production of humus (Fig. 5.13). In arid regions, under the influence of low soil moisture content the organic parts are mineralized first, however, the percentage of organic parts is very low (less than one percent). Also the peat soils of the tropics have a low decomposition rate of organic matter.

Some authors have estimated that the rise of the mean annual temperature by 10 °C accelerated the chemical weathering of soils threefold. Therefore, soils on landscapes of great age, such as presumably, the plateau surface of Brazil and Africa, have reached a very advanced stage under conditions not interrupted by glacials and interglacials.

In general, tropical soils in most parts of the tropics are infertile which seriously limits the development of tropical agriculture. Because of the rather low content of dissolved solids and suspended matter in tropical rivers these cannot create extensive and rich alluvial plains. According to various reports, even the Amazon lowlands or the Niger inland delta are not very suitable for crop production. Only those southeastern Asian rivers such as the Red River, the Mekong, the Menam, the Irrawaddy, the Brahmaputra, the Ganges and the Indus have a substantial amount of fertilizing particles and are capable of creating fertile plains and deltas. However somewhat less typically they flow from recent fold mountains of the Alpine age. Also fertile are the volcanic soils on slopes of tropical active volcanoes as in Indonesia.

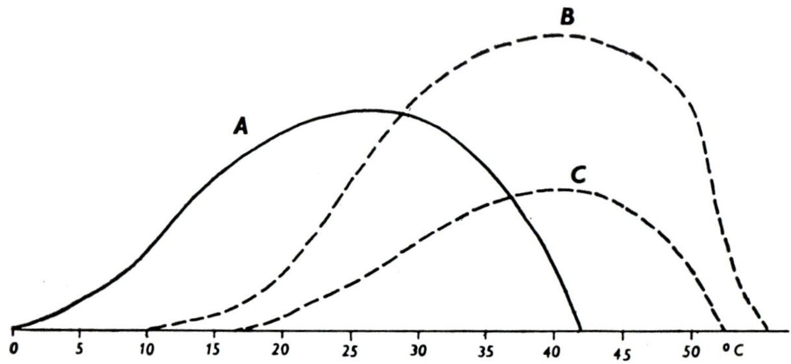

Fig. 5.13 Relationship between humus and air temperature, according to Kutílek. A, humus production; B, humus decomposition in humid environment; C, under water

The term fertility is relative. Gourou provides an example from Tonkin delta, where supposedly "relatively rich soil" contains less than $0.1^0/_{00}$ of potassium, while in temperate zones a soil with less than $0.5^0/_{00}$ is considered to be uncultivable. The same sample contained less than $0.1^0/_{00}$ of phosphates, while the minimum for cultivable soils is considered to be $0.2^0/_{00}$.

Tropical as well as volcanic soils, are more suited to agricultural production the younger they are due to the shorter periods of leaching to which they have been exposed.

Water exchange between soil, atmosphere and vegetation is a very complex process and soil and vegetation are subordinate factors. Puri [37] studied the relationship between soil and vegetation in various parts of India and thus on a large scale. He found subtropical evergreens south of Bombay on the red soils formed on granite, gneiss and laterite. The soils were poor in lime, magnesium. phosphorus, and nitrate, but rich in potash and contained very little humus. Typical for the Deccan were black soils developed on various substrata. The basaltic soils near Katray and Singraga were black in colour, fine grained and clayey, with a high content of iron, calcium, magnesium and aluminium but poor in phosphorus and nitrogen. The vegetation typical for such a type of soil was the deciduous forest with teak, *Terminalia, Anogeissus, Boswellia and Lannea.*

Formations of scrubby acacia were more typical for murrum soils while red soils with ferruginious clay bore an evergreen vegetation with species of *Eugenia, Actinodaphne, Terminalia bellerica, Mango,* etc. However, according to the author, in the wet tropics the evergreen vegetation on the ferruginious soil may differ considerably, which indicates that the type of soil itself is not the dominant factor in the ecological structure.

Greish to brown soils formed on sandstone, limestone and marble support a mixed forest with evergreen species of *Ficus* and *Eugenia.*

Such a type of study has been performed elsewhere, however, the conclusions can be used for further generalisation only when based on large scale observations.

Another effect of vegetational changes on the soil structure has been found mainly in relation to erosional processes. Truncated soil profiles under special conditions may appear on the surface as chemically different from the original soil profile and this may result in yet another vegetational change. This can also explain why the felling of evergreen types allows only the growth of moist deciduous or deciduous types of vegetation. Similarly, where deciduous species have been felled, they have been replaced by either savanna or by scrub with large patches of grassland. This has been attributed to a rapid decrease of soil nitrogen, phosphorus and organic matter.

A very valuable numerical evaluation of such changes was reported by Van Baven from various parts of the tropics [42]. For instance in Bolivia, where virgin forest had been replaced by sugar-cane a heavy drop in calcium, nitrogen and potassium was observed after the first year which still continued after the following three years. In the intervening time the amount of aluminium and manganese increased. Tab. 5.6 provides an example of such an evaluation. After several years of such a type of experiment Van Baren came to the belief that there was no possibility of arriving at a standard conclusion on changes in soil fertility and that for each particular type of soil and type of location a special analysis of soil in a virgin state and under cultivation should be performed.

In India, where the problem of soil cultivation is of vital importance and in keeping with the previous conclusion, specialised Soil and Water Conservation Branches were established on prevailing types of soil, namely: in Udagiri for eastern lateristic soils, in Jodhpur for desert and arid soils, in Hyderabad for red soils, in Agra, Kota and Vasad for ravine problems, in Bellary for semi-arid black soils and further centres for mountainous soils (Tejwani [41]).

In the process of water exchange between soil and vegetation the role of single elements have been studied, but with rather varied results. As far as nitrogen is concerned no shortage has been observed after many years of cropping on formerly forest soils but it is felt that the history of cropping can play a significant role. Many

Tab. 5.6 Comparative analysis of a virgin forest soil in Bolivia and the same soil under sugar cane, according to Van Baven

Aspects	Virgin forest soil	Same after 1 year : yield 60 tons of cane	Same after 3 years : yield 40 tons of cane
pH—H_2O	6.2	6.5	5.0
pH—KCl	5.4	5.6	3.9
Carbon %	1.1	0.5	0.2
Nitrogen %	0.125	0.068	0.051
P_2O_5 ppm/soil	30	tr.	tr.
Calcium	750	200	25
Magnesium	144	32	14
Potassium	69	33	15
Ammonium	5.5	5	8
Nitrates	4.0	1	11.5
Iron	tr.	2	6.5
Aluminium	2	8	50
Manganese	8	tr.	19
Sulphate	tr.	tr.	tr.
Sodium	81	20	19

tropical soils are not friable and thus do not have a high phosphorus absorption capacity. On lighter textured soils high residual effects were achieved from small applications. After two years of cropping on formerly forest soils deficiency in potassium was frequently observed. The deficiency was rarely found on savanna soils even if intensively cropped. A slow decrease in calcium and pH was observed after continuous cropping on forest and savanna soils. On soils where grass was regularly burned an increase of sulphur was observed while lower values of sulphur were found following the fallow.

Typical savanna soil is formed by a permeable layer superimposed on impermeable or permeable soil so that no true groundwater level exists. Some savanna soils are derived from moist forest soils by degradation. Poor sandy soils store water best, although they are often more favourable for degradation (Keag [25]). Also silty soils which are frequently wetter than sandy soils can be easily degraded when natural drainage is poor.

A lower content of exchangeable nitrogen and phosphoric acid provides conditions favourable for savanna and vice versa. A higher content of nitrogen and phosphoric acid is more favourable for forest. In addition, tropical forest soils contain more organic matter in their upper 50 – 100 mm than savanna soils. It is estimated that about 4 tons per acre of litter are added annually to the organic matter of forest soils.

An estimation of the nutrient content of west African soils (Nye [33]) is shown in Tab. 5.7.

Tab. 5.7 Content of nutrients in west African soils compared with forest as 100%, according to Nye

Nutrient	Forest 40 years old	Wooded savanna Grass	Woodland
P	100	1	2
K	100	6	18
Ca	100	1	10
Mg	100	7	18

Because of excessive heat even flooded soils and soils on volcanic slopes can be found to be less productive. Partial improvement can be achieved by planting species with large foliage, protecting the soil against the effects of heat. The long-term effect of measures of this kind is always very uncertain, particularly on a large scale, while when carried out on only a small scale the results may not be sufficiently convincing.

Extensive areas of soil along the lower reaches of the Amazon are under water from five to seven months of the year, the depth of the water being over 4 metres. In Pará region canals have been constructed 5 – 8 metres deep in an attempt to connect the Amazon main channel with Lake Maicum and build up the soil by siltation. There is estimated to be a potential daily transportation of three million tons of sediment and fifty thousand tons of potassium chlorate. The replantation of original and induced soils is found to be far from effective. What is known as Fordlandia in another part of the Amazon basin, planted with rubber trees was later changed to elephant and Guinea grass.

For preliminary soil studies the population density can serve as an indicator of the quality of the soil. For instance, in the Amazon basin the density of population is 1 to 2 km^2, while on Java it is 1000 to 1 km^2.

The process of soil water movement and soil water storage is often studied as a purely physical process on the basis of samples of soils processed in laboratories. However, there are many additional factors which affect water transportation through the soil profile which call for techniques applied in situ. For instance, the soil infiltration process is studied as a relationship between rainfall intensity and the infiltration capacity of the soil. This applies only under special conditions of permanently bare soils in semidesert and desert regions. The influence of the vegetation and its root system, not involved in the laboratory soil samples, may result in different infiltration rates than those achieved by using infiltrometers. The variability of evapotranspiration in the course of a year creates favourable conditions for soil saturation and makes water transport to the groundwater level highly variable.

Also the physical properties of soils such as wilting point or field capacity need to be carefully re-examined in the tropics. Soil moisture tension of about 15 atm. is widely accepted as what is known as wilting point, however, evidence exists that a number of plants, particularly in arid regions, may reduce the moisture content well below this level (Jackson [24]). The water potentially available to plants between field capacity and wilting point may not be actually available for various reasons related to the distribution of soil properties, especially soil texture, hydraulic conductivity, rooting depth and process of soil evaporation.

Although on special maps extensive areas of a particular soil type are to be found it has to be borne in mind that one type of soil a high variability of soil water characteristics and hydraulic conductivity can be found, as has been experimentally proved by Babalola [3] for Nigerian soils.

Mohr and Van Baven [32] recognized the following types of prevailing vertical movement in tropical soils:

a) A continuous downward movement in regions with no dry season. The permanent inflow of water into a wetted soil profile results in continuous leaching.

b) Alternating downward movement and cessation is typical for wet and dry regions where the groundwater level is at such a depth that the capillary rise cannot transport water to the surface and the upper layers dry out seasonally. Leaching occurs only during the wet season, otherwise slight rains will evaporate from the upper layers of the soil.

c) Ascending water movement will occur when there is very low rainfall and high evaporation from continuous transport of capillary water enriched by salts. The process results in the accumulation of salts at the surface.

d) Alternating upward and downward movement occurs in wet and dry regions when capillary water can reach the surface layers and evaporate. The surface soil layers are saturated from the groundwater during the dry season and the alternating process of leaching and washing occurs according to the balance between the rainfall, groundwater and evaporation regimes.

e) Intermittent ascending movement will occur in regions with a prolonged dry season when the groundwater level is so deep that it transports salts only for a short period of the year, otherwise the process is stopped by temporarily wetted upper soil layers.

Special conditions may produce soils with low permeability or desert soils not included in the above classificaton.

Fig. 5.14 traces the major types of soil found in the tropics. *Laterites and latosolic soils* are regarded as a single group. Those soils which have undergone lateritic processes are characterized as lateritic soils (Fig. 5.14A). Yellow soils, red soils and laterites of the humid tropics belong to this group.Yellow and red soils are sometimes called latosols and a typical feature of them is leaching and migration of SiO_2. All soils are heavily leached of salts with an accumulation of iron, aluminium, manganese and silica. Because of the rapidity of bacterial activity, humus is lacking. The soils are mostly reddish in colour and soft when covered but become hard when exposed to heat. The soils are poor, acid and lacking in mineral nutrients. Oxides are accumulated to such an extent that compact layers known as laterites are often formed. This occurs in regions with equatorial climates and savannas.

Typical soils in this category are sawah in Indonesia, nipe clay and matanzas in Cuba, tanah merah in Indonesia, ouklip soil in South Africa, brown earth in Trinidad and paddy soil in China.

From the hydrological point of view good drainage is necessary if lateritic processes are to occur. Saturation of the upper layers through the capillary rise makes possible the transportation of various iron compounds to the surface, however, such a type of process is not only typical of laterites. Some laterites become very granulated in the upper layers so that the contact of their roots with the capillary pores is rather limited.

Podzols (5.14B) are formed in an acid medium and are characterized by the destruction of silicates, particularly in the clayey fraction. Released sequioxides migrate in the soil profile. Low condensation of humus is another feature of tropical podzols. The process of podzolization may occur in tropical lowlands, although it is more frequent in mountainous regions. A whitening of the layers below the rough humus layer is typical of the profile. Typical podzols can be found in the mountainous parts of Tanzania and in New Guinea.

Margalitic soils can be found under semi-humid conditions, also occasionally located in the humid and semi-arid tropics. Dark grey and black in colour they are sometimes classified as chernozems, although in contrast to real chernozems they have a humus low content. $CaCO_3$ is found in all layers of the soil profile, but the carbonate layer is missing. A typical parent rock is limestone; the *chernozems* are also associated with dark-coloured igneous rocks. (Fig. 5.14C). They tend to develop on flat plains. The black layer is less than 10 cm thick, is rich in humus and fades into a yellow-brown horizon, which is divided by a sharp line from a light coloured horizon. During heavy rains they are sticky and heavy while during dry seasons they become cracked.Margals are typically found in Ethiopia and Indonesia, regur in India, Ethiopia, Sudan and southern and western Africa, chaga in Zanzibar, dambo in Zambia and Zimbabwe, mkuzi in southern Africa, black cotton soil in India, marlsoil in Indonesia and mallee soil in Australia.

Another group is formed of *red loams, chestnut* and *brown soils* and are typical in semi-arid areas with low annual rainfall. Red loams also develop on rolling land with a rainfall of over 1000 mm and a high temperature. The parent rock is granite, schist, sandstone loess and loamy alluvial sediments (Fig. 5.14D). Chestnut and brown soils under semi-arid conditions have a lower content of humus but a high content of humic acid. The salts are often transported to the surface as a result of the capillary rise. They are typical of the tropical savannas.

Desert soils (5.14E) have a grey or red colour and contain little humus because of the limited vegetation. Horizons are slightly differentiated. The profile tends to be shallow due to the absence of leaching by rainfall. Calcium carbonate occurs in the form of a lime crust, appearing as a hard rock layer due to the slow evaporation of water near the surface. In depressions where there is no outlet for intemittent streams, evaporation is more intensive and the upper layers contain more salts. Some of the soils are aeolian in origin and thus bear little relation to the rock on which they rest.

Sometimes saline and alkaline soils in arid and semi-arid regions are considered as a separate group of *solonchaks*. They are also known as reh and usar in India and as tanah assinan soils in Indonesia. A basis feature of these soils (Fig. 5.14F) is the vertical upward movement of salts in the soil profile under the impact of evaporation which exceeds rainfall. This indicates that they are not only related to the semiarid and arid tropics. Most frequent are salts of anionts CO_3^{2-}, HCO_3^-, Cl^-, SO_4^{2-} and kationts Ca^{2+}, Mg^{2+}, K^+, Na^+. The salts are accumulated at a certain level balanced for the given hydrologic conditions. With decreasing rainfall and increasing evaporation the layer moves upward and under extreme conditions a white colour predominates on the soil surface. This process can be accelerated by uncontrolled irrigation.

Tab. 5.8 Shifting of the layer enriched by salts, dependent on rainfall and temperature, according to Kutílek [26]

Soil	Annual rainfall (mm)	Mean annual temperature (°C)	Depth of salt concentration (mm)
Chestnut	200—350	3—8	350
Grey	100—350	13—16	100—200
Solonchak	100	over 16	surface

The shifting of the layer enriched by salts is given in Tab. 5.8. In general, towards the equator the groundwater and soil water becomes increasingly mineralized. If the groundwater is in the vicinity of the soil surface, the probability of salinization through the capillary rise increases. The salinization process is influenced by the hydrological and biochemical regimes as well as latitude.

5.8 SOILS IN THE TROPICS

Six regions can be recognized in tropical Africa. In the Sahara desert soils can be found, although in many parts there are sandy areas and moving dunes or another aeolic type called hammada. Depressions are filled with fine textured loamy silts. In southern parts of the Sahara soils gradually change into the form of chestnut soils. Laterisation processes can be traced everywhere in the Sahara, except for the oases. The Nile valley has alluvial soils, partly sandy and partly clayey, frequently with a high content of salts.

In the Sudan the effects of humidity increase. North of Lake Chad red-cinamon soils are to be found transferred southward into chestnut and margalitic soils. Solonchak occurs sporadically. Margalitic soils prevail in Ethiopia and large parts of the Sudan. In arid parts of east Africa brown and red-brown soils prevail, mixed with greyish desert soils.

In equatorial Africa there are extensive areas of lateritic soils. Alluvial soils are found along great rivers such as the Congo. Exception in the equatorial part of Africa are the fertile juvenile soils on the slopes of great volcanoes.

Various soils are found in the region of great African lakes. A highly variable orographic pattern, rainfall distribution and climate form a mosaic of chestnut, cinamon and margalitic soils. Podzols, black soils and laterites exist also in hilly regions. In Malagasy lateritic soils and chestnut soils are dominant; also terra rosa and grey desert soils are found in some smaller regions. Red loams are found in Mozambique.

South African soils are under the increasing impact of aridity. Desert soils are found in the Kalahari together with solonchaks. In more humid parts there are chestnut soils and yellow soils and in mountainous parts podzols are frequent.

A great variety of soils are to be found in Latin America. In the eastern highlands of Guyana and Brazil there are red soils in various forms depending on the parent rock, climate and vegetation. Terra rosa is a typical soil here and is rich in humus content. Podzols are also typical of the region. In southern Brazil the red soils are transformed into black soils which in general are very fertile.

Extensive regions of the Amazon lowlands and Orinoco delta are formed of alluvial deposits. The soils are swampy. In Gran Chaco soils are margalitic alluvial formations rich in humus. A great variety of soils are to be found in Bolivia; in the eastern parts there are tropical laterites, further south with decreasing humidity margalitic soils are more common. The most fertile soils are outside the tropical limits.

In western mountainous regions alpine types of soil are found, with a high humus content. This is particularly so in Bolivia, Ecuador, Colombia and Peru. Podzols are found below the forest line, these being transformed at lower altitudes to yellow and red soils. In semi-arid and arid regions there are greyish desert soils and solonchaks.

In the Caribbean region, clayish soils called matanzas can be found in Cuba. Terra rosa is also widespread. In flat regions solonchaks are found. Bayamo types of soil on limestone rock are less fertile.

Most significant for India are margalitic soils, known as black cotton soils; in the southern and eastern parts there are red and yellow soils, along the west coast are laterites. In Sri Lanka there are laterites and terra rosa. Desert soils have a high content of salts.

In volcanic regions, such as those of Indonesia there are basalt soils and volcanic sediments. Most of the soils are juvenile, brown and yellow in colour, parts are senile in a lateritic stage. Less fertile sawah soils are a product of frequent artificial flooding.

5.9 REFERENCES

[1] Airey, P. C., Calf, G. E., Campbell, B. C., Harley, P. E., Roman, D., Habermehl, M. A., 1979. Aspects of the isotope hydrology of the Great Basin, Australia. Istotope Hydrology, IAEA Vienna, 205—219.

[2] Athavale, R. N., Murti, C. S., Chand, R., 1980. Estimation of recharge to the phreatic auqifers of the Lower Maner Basin, India, by using the tritium injection method. Journal of Hydrology, 45, 185—202.

[3] Bababola, O., 1978. Spatial varability of soil water properties in tropical soils of Nigeria. Soil Sc., Vol. 126, No. 5, 1978, 269—278.

[4] Balek, J., 1979. Conceptual simulation of groundwater storage fluctuation in extensive sedimentary structures. IAH Memoirs, Vol. 15, Vilnjus.

[5] Balek, J., 1981. Unesco India postgraduate hydrological education and research. Unesco Rep. IND/78/056, New Delhi, 8 p.

[6] Baweja, B. K., 1980. Groundwater development in India. Water Qual. Bull, Vol. 5, No. 4, 82—83.

[7] Bose, R. N., Ramakrishna, F. S., 1978. Electrical resistivity survey for groundwater in Deccan trap country of Sanghi district, Maharashtra. Journal of Hydrology 38, 209—222.

[8] Burgman, J. O. S., Erikson, E., Kostov, L., Möller, A., 1979. Application of oxygen-18 and deuterium for investigating the origin of groundwater in connection with a dam project in Zambia. Isotope Hydrology IAEA Vienna, Vol. 1, 27—42.

[9] Calf, G., E., 1978. The isotope hydrology of the Mereenie Sandstone aquifer, Alice Springs, N. Territory. Journal of Hydrology 38, 341—356.

[10] Dijon, R., 1981. Groundwater management. Water Qual. Bull., Vol. 5, No. 4, 77—81.

[11] Dixey, F., 1962. Geology and geomorphology and groundwater hydrology. Unesco Symp. on the Prob. of Arid Zones, Paris.

[12] Everett, L. A., 1981. A structured groundwater quality monitoring for developing countries. Water Qual. Bull. Vol. 6, No. 1, 1—6.

13] Fritz, P., Suzuki, O., Silva, C., Salati, E., 1981. Isotope hydrology of groundwater in the pampa del Tamarugal, Chile. Journal of Hydrology 53, 161—184.

[14] Geygh, M. A., Wirth, K. W., 1980. ^{14}C ages of confined groundwater from the Gwandu aquifer, Sokoto basin, Northern Nigeria. Journal of Hydrology 48, 281—288.

[15] Gonzales, T. J. A., 1981. A strategy for management of groundwater in Mexico. Water Qual. Bull. Vol. 6, No 1., 19—21.

[16] Gustaffson, Y., 1979. Groundwater resources development experiences for the coimbatore project in southern India. In "Hydrology in Dev. Countries, Ed. Riise and Skofteland", Nordic IHP Rep. No. 2, 257—260.

[17] Haman, Z., 1979. Importance of groundwater data bank for an optimum planning of groundwater development. In "Hydrology in Dev. Countries, Ed. Riise and Skofteland", Nordic IHP Rep. No. 2, 269—275.

[18] Heaton, P. H. E., Vogel, J. C., 1981. "Excess air" in groundwater. Journal of Hydrology 50. 201—216.

[19] Hellwig, D. H. R., 1973. Evaporation of water from sand. Journal of Hydrology 18, 93—108.

[20] Hellwig, D. H. R., 1978. Evaporation of water from sand. Journal of Hydrology 39, 129—138.

[21] Hills, E. S., 1953. Hydrology of arid and subarid Australia with special reference to underground water. In Reviews of Research on Arid Zone Hydrology, Unesco, Paris.

[22] Hughes, R. J., 1981. Groundwater quality in Australia. Water Qual. Bull., Vol. 6, No. 1, 15—18.

[23] Jacks, G., 1982. Nitrogen circulation and nitrate in groundwater in an agricultural catchment in southern India. Int. Symp. on Groundwater Pollution from Agric. Activities, Prague

[24] Jackson, I. J., 1977 Climate, water and agriculture in the tropics. Longman, London, 248 p.

[25] Keag, R. W. J., 1959. Derived savanna-derived from what? Bull. de l'IFAN 21, 427—438.

[26] Kutílek, M., 1963. Pedology of the tropics and subtropics. (In Czech). SPN Prague, 107 p.

[27] Lloyd, J. W., 1976. The hydrology and water supply problems in north-central Chile. Pacific Science, Vol. 30, No. 1, 9—19.

[28] Mann, A. W., Deutscher, R. L., 1978. Hydrochemistry of a calcrete-containing aquifer near Lake Way, W. Australia. Journal of Hydrology 38, 357—378.

[29] Martin, H., 1961. Hydrology and water balance of some regions covered by Kalahari sands in S. W. Africa. Symp. on Afr. Hydrol. Proc., Nairobi, 450—455.

[30] Mazor, E., Verhagen, B. Th., Sellschop, J. P. F., Jones, M. T., Robins, N. E., Hutton, L., Jennings, C. M. H., 1977. Northern Kalahari groundwaters. Journal of Hydrology 34, 203—234.

[31] McCann, D., L., 1971. The role of groundwater discharge in the water balance of Lakes Victoria, Kyoga and Albert. Interim UNDP Rep., Entebbe, 38 p.

[32] Mohr, E. C. J., Van Baven, F. A., 1972. Tropical soils. Interscience Publishers Ltd., Amsterdam.

[32] Nace, R. C., 1969. Human use of groundwater. In Water, Earth and Man", Methuen and Comp., London.

[33] Nye, P. N., 1959. Some effects of natural vegetation on the soils of West Africa. Trop. Soils and Veget. Symp. Proc., Abidjan. Unesco, Paris, 54—67.

[34] Pearson, F. J., Swarenski, N. W., 1974. ^{14}C evidence for the origin of arid region groundwater, N. E. Province, Kenya. Symp. on Isot. Tech. in Groundw. Hydrol. Proc., IAEA, Vienna, 1—95.

[35] Piancharoen, Ch., Soitrahul, S., 1980. Groundwater legislation and management in Thailand. Water Qual. Bull, Vol. 5, No. 4, 86—88.

[36] Prichett, A., 1975. Case history of groundwater development in the Zapotitan Valley, Republic of San Salvador. Adv. in Hydroscience, Vol. 10, 86—96.

[37] Puri, G. S., 1959. Vegetation and soil in tropical and subtropical India. Tropical Soils and Vegetat. Symp. Proc., Unesco, Paris, 93—102.

[38] Roose, E. J., Lelong, F., 1981. Factors of the chemical composition and groundwater in the ITCZ, West Africa. Journal of Hydrology 54, 1—22.

[39] Shyakubov, B., Morales, L., 1980. Consideraciones para la exploatacion de acuiferos de nitrusion of marina. Voluntad hidraulica, Año XVII, 10—15.

[40] Taylor, J., 1973. An atlas of Pacific islands rainfall. Document HIG, Hawai, Institute of Geophysics, Honolulu.

[41] Tejwani, K. G., 1970. Soil and water conservation — promise and performance. Indian Journ, Soil Conserv., Vol. 7, N. 2, 80—86.

[42] Van Baren, F. A., 1959. The pedological aspects of the reclamation of tropical and particularly volcanic soils in humid regions. Trop. Soil and Veget. Proc., Unesco, Paris, 65—67.

[43] Verhagen, B. Th., Smith, P. E., McGeorge, I., Dziembowski, Z., 1979. Tritium profiles ni Kalahari sands as a measure of rain water recharge. Istotope Hydrology, IAEA Vienna, 733—751.

[44] Vogel, J. C., et al., 1963. A survey of the natural isotopes of water in South Africa. Radioisotope in Hydrol. Symp. Proceedings, Tokyo. IAEA Vienna, 407 p.

[45] Vogel, J. C., Talma, A. S., Heaton, T. H. E., 1981. Gaseous nitrogen as evidence for denitrification in groundwater. Journal of Hydrology 50, 191—200.

[46] Williams, A. F., Holmes, J. W., 1978. A novel method of estimating the discharge of water from mound springs of the Great Artesian Basin. Journal of Hydrology 38, 261—272.

[47] Anon., 1974. World water balance and water resources of the earth. (In Russian). Gidrometeoizdat, Leningrad, 638 p.

[48] Young, A., 1974. Some aspects of tropical soils. Geography 1974, 233—239.

[49] Yurtsever, Y., Payne, B. R., 1979. Application of environmental isotopes to groundwater investigation in Quatar. Isotope Hydrology, IAEA Vienna, 465—484.

[50] Zafiryadis, J. P., 1979. Hydrogeological investigation to improve the rural water supply situation, case study from Niger. In "Hydrol. in Dev. Countries, Ed. Riise and Skofteland", Nordic IHP Rep. No. 2, 291—307.

6 LAKES AND SWAMPS

6.1 THE RELATIONSHIP BETWEEN LAKES AND SWAMPS

Lakes and swamps form an important part of tropical water resources. They contribute significantly to the water economy of many tropical countries acting as storage reservoirs and centres of the fishing industry. Since ancient times they have been one of the main sources of food and water and the favourable conditions along their shores has been one of the positive factors in the development of tropical civilizations. New artificial lakes have been formed during recent decades and still more are under construction. Some of them are among the largest man-made lakes in the world. Only in the recent past has more attention been paid to the construction of small shallow reservoirs which seem to be more effective particularly for the purpose of commercial fisheries. Besides this they can be constructed and maintained by means of self-help schemes.

Swamps occupy much larger tropical areas than lakes. However, while the characteristics of natural and artificial lakes are well known, much less is known about the morphological and hydrological characteristics of tropical swamps. An account of them was given for Africa by Balek [3], where the total size was estimated at 340 000 km^2. It is to be expected that their area in South America and Asia is considerably greater. In Sumatra alone, the area of coastal mangrove swamps is estimated to be 150 000 km^2, about one-third of the total size of the island.

A significant portion of the swamps consists of headwater swamps smaller in size as single units, but just as important as the large ones.

A negative effect of lakes and swamps on the hydrological balance is water loss resulting from evaporation from the free water surface and evapotranspiration from the aquatic vegetation. A simple scheme gives an idea of the relationship between lakes and swamps in the tropics:

Lake/pond – perennial swamp – intermittent swamp – land.

Headwater area/lake – perennial headwater swamp – intermittent headwater swamp – land.

In contrast to lakes, swamps are clearly related to the vegetation. According to Welch [17] swamps belong genetically to the standing water series in which the water motion is not that of a continuous flow in a definite direction, although a certain amount of water flow may occur as internal currents in the vicinity of the inlets and outlets. What are regarded as lakes are inland bodies of standing water of a large size. According to the World Meteorological Organization glossary, swamps, marches and bogs are described jointly as lowlands flooded in the rainy season and usually watery at all times. Specialists in swamp hydrology differentiate between swamps and marshes on the one hand and bogs on the other. Bogs are characterized as being genetically related to lakes as a possible final stage of lake development, while marshes and swamps are defined as land area covered by vegetation and saturated with water.

Obviously the terminology is still far from being uniform, so ecological and hydrological evaluation should be a part of any study concerned with swamps, marshes and bogs.

The eutrophication process has been found to be vitally important in the global water economy of the standing water series.

Barabas [4] stated that although eutrophication problems have been most acute in lakes and reservoirs, they also arise and must be dealt with in running waters, particularly in estuaries and coastal areas. Eutrophication has been described as the process of aging of lakes, but according to Barabas, such a description emphasizes the effects rather than the causes.

In the tropics particularly lakes are only a temporary feature of the surface of our planet. Most tropical lakes have been formed by the actions of rivers and wind and by earth movement and tectonic activity. They disappear from the earth through a process of natural eutrophication involving the filling of lakes with nutrient-containing sediments. The process of cultural or man-made eutrophication is much faster than the slow rate of natural eutrophication. Among these nutrients, many of them beyond any control, the most frequent are carbon, hydrogen, oxygen, nitrogen, phosphorus, sulphur, potassium, magnesium and calcium.

It has been proved that phosphorus content is a determining factor of eutrophication and the reduction of phosphate loadings and the removal of phosphates already present in lakes is one of the most effective means of eutrophication control. In addition to this the removal of organic matter before decomposition and the introduction of herbivorous fish are another effective means of control.

So far the most effective eutrophication studies in semi-arid areas have been undertaken in Australia. Cullen and Smalls [5] proved that the Australian lakes do not behave in the same way as the cool lakes of temperate regions. The great variability of rainfall in a single year has a negative effect on lake regimes. Semi-arid soils are highly erodable and if the vegetative cover is damaged by overgrazing or drought then a large volume of sediments enriched with superphosphates enters waterways and lakes. The flow rate into the lakes is more important than concentration in controlling the export of nutrients, as can be seen in Tab. 6.1.

Eutrophication can have a considerable influence on the hydrological regime of lakes and river outflows. The formation and destruction of vegetational barriers at the outflow point of Lake Nyasa results in the fluctuation of the level of the lake. The effect of growing vegetation is later combined with the effect of animals.

Tab. 6.1 Phosphorus exports to Lake Burley Griffin as dependent on the flow regime. According to Cullen and Smalls [5]

Flow regime	Duration (days)	Total flow ($\times 10^6$ m^3)	Mean P. concentration (mg P. m^{-3}) Total P.	Total diss. P.	Exports (kg)
Normal	354	142.4	28	15	3900
Drought	159	31.5	17	6	540
Flood 1975	37	167.4	125	24	21 000
1976	13	174.7	257	26	44 800

Beside artificially induced herbivorous fish the hippopotamus can influence swamp morphology as it grazes on the swamp vegetation.

In principle there are two concepts of the formation of swamps. Welch [17] related swamps to the so-called static environment or to the standing water series. In contrast, Debenham [6] suggested the vegetational concept based on the principle of running water, assuming that aquatic vegetation requires water that flows. Such an assumption resulted in the conception of the inclined surface and the conclusion was that the life span of swamps is longer than that of lakes, because the vegetation of swamps can adapt itself to many physical changes apart from continuous drought.

Apparently both concepts are valid under certain circumstances, one can imagine an intermittent swamp, even if the intermittency is related to a period of more than one year. Very little or no aquatic vegetation is found in areas flooded once in several years, while in the seasonally inundated areas the conditions for aquatic vegetation are very favourable.

It should be remembered, however, that both concepts were established long before the existence of a pronounced impact by man on the lake/swamp regime.

6.2 TROPICAL LAKES

Tab. 6.2 lists the world's largest natural tropical lakes. The largest lakes are mainly in Africa, there are fewer in South and Central America and only one significant lake in the tropical areas of Asia. Great lakes in Australia are outside tropical limits.

The greatest African lake and the third largest in the world is Lake Victoria, also known as Ukereve. It is more than 400 km long, 240 km wide and according to some sources it covers an area of more than 69 000 km^2. The size, however, depends on whether the coastal swamps along the lakeshore are regarded as part of the land or of the lake. The shore is 7000 km long. Geologists assume that the lake

Tab. 6.2 Largest lakes in the tropics

Lake	Country	Area (km^2)	H max (m)	Volume (km^3)	Altitude (m.s.l.)
Victoria	Tanzania, Kenya, Uganda	66 400	92	2656	1135
Tanganyika	Zambia, Rwanda, Burundi, Tanzania, Zaire	32 890	1435	18 940	773
Nyasa	Tanzania, Mozambique, Malawi	30 800	706	7000	472
Chad	Chad, Niger, Nigeria	18 000	12	27	240
Maracaibo	Venezuela	13 000	250	—	0
Tonlé Sap	Kampuchea	10 000	12	40	15
Rudolph	Kenya	8660	72	—	427
Nicaragua	Nicaragua	8430	77	108	32
Titicaca	Peru, Bolivia	8110	400	710	4100
Albert	Uganda, Zaire	5300	57	64	620

once covered a much larger area than it does today and included Lake Kyoga. This means that the water level was at least 90 metres higher than today. The seasonal fluctuation of the lake level averages 65 cm, however, occasionally, for example in the 1960s, higher fluctuations have been observed.

Lake Tanganyika is the seventh largest in the world. Since the maximum depth of the lake is 1435 metres, the lake bottom is far below sea level. In fact Tanganyika is the second deepest lake in the world, after Baikal. A subsurface ridge divides the lake into two pans. The only outflow from the lake, the river Lukuga, is intermittent and lake evaporation often exceeds precipitation and inflow. An outlet from the lake was made artificially in 1878 by digging through the banks at Albertville and the lake level went down by 10 metres. The temperature in the first 480 metres varies between 25 – 27 °C and from this depth it is fairly constant at 23.1 °C.

Lake Nyasa is a very long lake and like Lake Tanganyika it is still in the process of tectonic development. The basin of the inflows measures about 100 000 km^2 and the lake is drained through the River Shire into the Zambezi. Seasonal fluctuation is about 1 metre.

Lake Chad is located in a shallow depression and owing to the flat relief the lake area varies greatly depending on the regime of the main inflows – Logone and Chari. Thus its size can vary between 10 000 and 25 000 km^2, while the mean depth is only 150 cm. The greatest stage occurs in July when the lake length increases to 250 km and the width to 150 km. The mean annual fluctuation is below 80 cm, the maximum being 200 cm. Because both inflows carry a great amount of sediment, the water becomes brackish. In a northeasterly direction the lake is intermittently connected by a surface runoff with the Bodel pan.

Maracaibo is a lagoon lake in Venezuela which fills a tectonic depression near Tablazo Bay, connected with the ocean by a short shallow channel. In the southern parts the lake water is fresh, in the northern parts it is brackish. The salinity varies between $400-2300\ \mathrm{mg\,l^{-1}}$, depending on the rate of exchange of water between the lake and the Caribbean Sea and on the hydrological regime of the lake inflows.

The only representative of Asian lakes Tonlé Sap is in the Mekong basin and stores excess flood water carried by the Mekong, the culmination of which is in September.

At the margins of the Great Rift formations, south of the Ethiopian Highland is Lake Rudolph, 300 km long and 25 – 60 km wide. The maximum depth is 70 metres. Most of the inflow is from the Omo River, there is no outflow from the lake.

Nicaragua is a lake which fills another depression formed by volcanic activity in Central America. The San Juan river drains the lake into the Caribbean Sea.

Tititaca is the highest of tropical lakes. It fills a tectonic depression and is a relict of an earlier, much larger lake. The lake drains through the River Desaguadero into Lake Poopo. There are several inflows into the lake, the largest being the Ramis. There is no outflow from the Poopo. The temperature at the surface fluctu-

Tab. 6.3 Largest artificial lakes in the tropics

Reservoir/ river	Country	Storage (km^3)	Area (km^2)	Dam height (m)	Finished
Kariba Zambezi	Zambia Zimbabwe	160	5250	100	1963
High Aswan Nile	Egypt Sudan	157	5120	95	1971
Volta Volta	Ghana	148	8480	70	1967
Itaipu Paraná	Brazil Paraguay	129	1400	190	1988
El Mantecho Caroni	Venezuela	111	—	136	1968
Pa Mong Mekong	Laos	107	—	115	1977
Cabora Bassa Zambezi	Mozambique	66	2700	100	1977

ates between 11 – 14 °C, deeper there is a constant temperature of 11 °C. The water is moderately saline.

The natural landscape of the tropics has been greatly changed during the past decades by the activity of man. The construction of dams changed the world of the tropics to a great extent and influenced ecological, economic and sociological life in extensive areas. Tab. 6.3 gives a list of the largest artificial reservoirs of the tropics. Even larger schemes have been planned in various parts of the tropics. The main problems do not relate to constructional aspects but to the possibility of predicting the impact of the new reservoir on life in the region.

More artificial lakes of a smaller size are to be found in the tropics, and, a description of them goes beyond the scope of this book. A list of them can be found in specialized publications [18].

Far less is known about the hydrological regime of lakes than of rivers. Tab. 6.4 shows the water balance values calculated for some of the African lakes. The balance of lakes is very sensitive to artificial influences and can easily be temporarily or permanently disturbed as in the case of Lake Tanganyika.

Fluctuation in the levels of lakes reflects the complexity of natural phenomena. By using geological and geomorphological indication of long-term level fluctuation, the hydrological regime can be traced back to geological history. Sometimes even short term fluctuations and sudden changes in the level of lakes indicate remarkable

Tab. 6.4 Water balance of African lakes related to lake area

Lake	Area (km^2)	Inflow (mm)	Precipitation (mm)	Outflow (mm)	Evaporation (mm)
Victoria	66 400	241	1476	316	1401
Tanganyika	32 890	1609	950	141	2418
Nyasa	30 800	472	2272	666	2078
Kariba	5250	8440	686	7038	2088

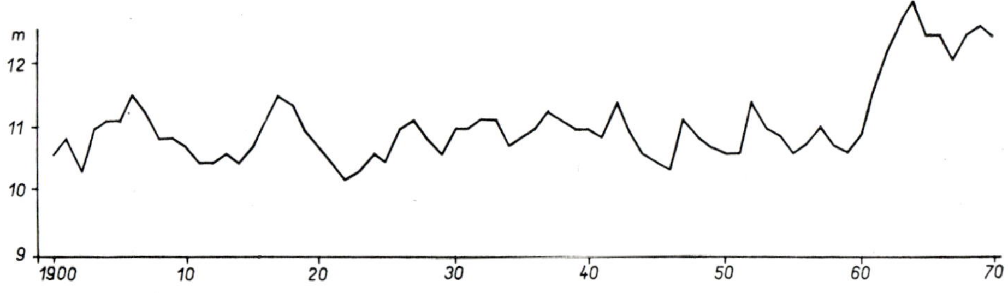

Fig. 6.1 The fluctuation of maximum levels of Lake Victoria

changes of various climatological factors. The sudden change of the level of Lake Victoria shown in Fig. 6.1 and the continuous rise of Lake Nyasa after 1915 shown in Fig. 6.2 provide examples of this.

Evaporation from lakes has only occasionally been systematically studied. For

Tab. 6.5 Hurst's measurement of evaporation from free water surface

	Oases	Khartoum	Lake Victoria	Lake Albert	Lake Edward
Evaporation (mm)	6500	7600	3900	3400	3900

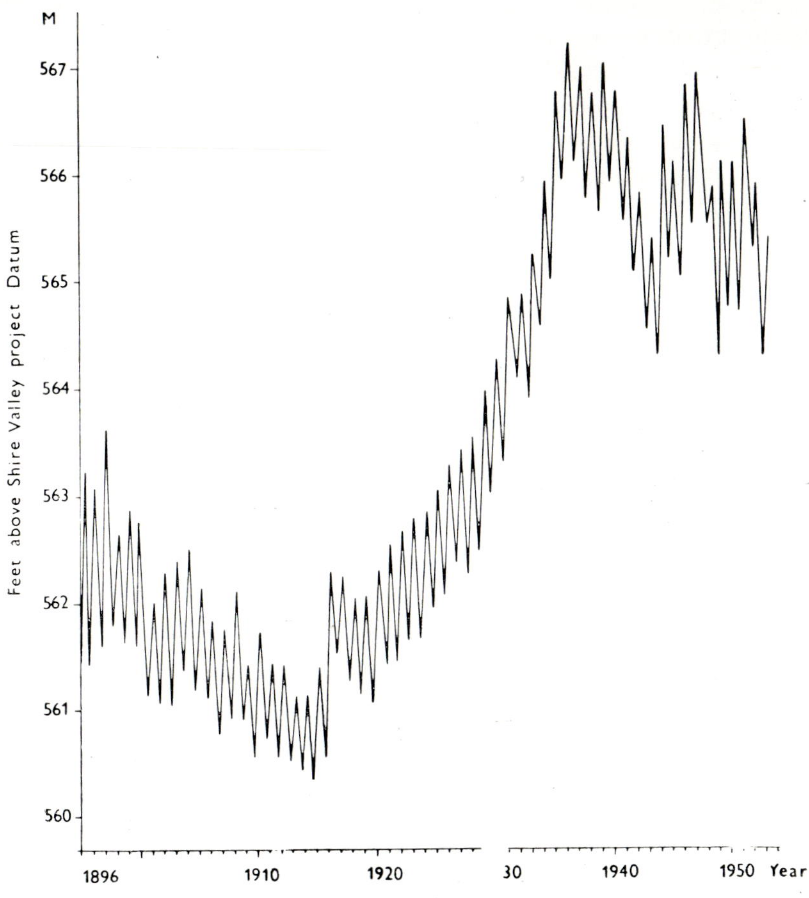

Fig. 6.2 The fluctuation of the levels of Lake Nyasa

instance, Hurst [9] gave an account of the results of measurements of evaporation from the open surface as shown in Tab. 6.5. A comparison of these values with the results obtained by the water balance calculation in Tab. 6.4 indicates considerable deviations. Modern water balance calculations from various sources indicate that evaporation from a free water surface in the tropics varies between 1250 and 2500 mm.

The saline content of lakes is highly variable, depending on the hydrological regime of the lake and the inflows and on the geological formations of the basin. For instance the salinity of great lakes in Africa varies between 65 ppm for Lake Victoria and 480 ppm for lake Mobutu. In general, lakes without surface drainage or with a substantial contribution from the baseflow have a high saline content.

6.3 TROPICAL SWAMPS

Swamps are areas with standing or slowly moving water in the non-capillary pores of the upper layers of the soil combined with vegetation-covered land over-saturated with water temporarily or permanently. Seasonal swamps cover an area of 250 000 km^2 in southwestern Brazil only and in Africa the total number of small intermittent swamps is estimated at between 10^4 and 10^5. South American intermittent swamps known as 'pantanal' result from about 1200 mm of rain falling during the rainy season on less permeable soil.

Tab. 6.6 Area of swamps in tropical parts of the continents

Continent	km^2
South America	1 200 000
Asia	350 000
Africa	340 000
Australia	2000

The hydrological characteristics of swamps are closely related to the degree of decomposition of plant residues. As can be seen from Tab. 6.6 the largest area covered by swamps of all types is to be found in South America where conditions are very favourable for the formation of swamps. In general, these conditions are:

a) flat relief and impermeable soil or rock close to the surface;

b) clearance of forest areas, so that the groundwater level rises near to the surface;

c) a relief pattern formed in such a way that it can absorb excess water from a much larger basin during the rainy season;

d) the formation of aquatic vegetation on slowly flowing streams or lakes;

e) margins of the river valleys or coastal regions where the groundwater aquifer is incapable of storing all the groundwater recharge.

Debenham [6] studied a variety of African swamps and defined them as products of those types of vegetational cover which tend to hold the backwater. Under such circumstances the morphological conditions appear to be secondary, although this is not always quite correct.

Kimble [12] found the following to be typical features of swamps:

a) runoff regulating systems acting in principle as reservoirs with an increased rate of evapotranspiration,

b) high ratio of surface area to water depth,

c) fluctuation in the size of swamps from year to year and in some cases from season to season,

d) three clearly marked zones: a zone at the margin of the swamp is under water for only a brief part of the year, a second zone which is waterlogged for a much longer period and a third zone which is under water throughout the year.

According to some authors seasonal and perennial swamps are of the same origin, both occurring whenever the morphological and climatological conditions result in the waters from the rainy season collecting in a locality faster than they can disperse. Thus under extreme conditions some swamps last only for several weeks while other intermittent swamps, as in the Zambezi basin, may become fully saturated and perennial for several years.

A significant difference in swamp morphology can be traced to the source of swamp recharge. Ordinary swamps are partly recharged by the precipitation falling on their surface, however streams normally serve as the main source. Headwater swamps known in Africa as dambos are formed in the upper sections of the channel network where erosion cuts across a valley where the rock is less weathered and covered with small deposits of soil. This type of swamp is recharged mainly by precipitation, since the subsurface inflow is relatively small and passes through them into the drainage network. The surface runoff into them is negligible. There the surface layer is thin but can still be regarded as a temporary shallow stream draining the whole area. A comparison of two types of swamps is given in Fig. 6.3. The drainage network in the second type can often be traceable only by remote sensing methods. The spongy effect of the headwater swamps (Balek Perry [2]) is highly beneficial to the hydrological regime in the lower reaches because they can retain excess water during the rainy season and release it during the months without significant rainfall. Obviously the regime of those swamps is very sensitive to any activity on the part of man and they can be easily damaged as a result of overgrazing and unprotected cultivation, after which the overland flow may result in sheet erosion. Also the ditches draining the swamps may reduce or damage any positive effect by accelerating surface runoff. The separation of the groundwater flow, surface flow and flow from the headwater swamp in the upper reaches of the Zambezi clearly indicates the delaying effect of the surface runoff (Fig. 6.4).

Obviously vegetation plays an important role in the hydrology of swamps and dambos. The genetic concept of swamps and their energy balance is related to the role of vegetation.

A simplified model of the energy balance of the swamps can be formulated as:

$$R = F + P + L + G + O$$

where R is the surface net radiation, F is the flow of heat into the aquatic biomass and soil, P is the energy spent on heating the air, L is the energy spent on evaporation, G is the energy spent on photosynthesis and O is the energy spent on evapotranspiration.

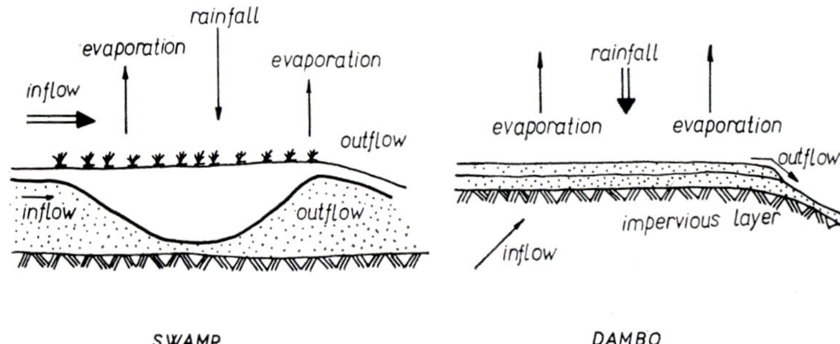

Fig. 6.3 Comparison of a perennial and headwater swamp

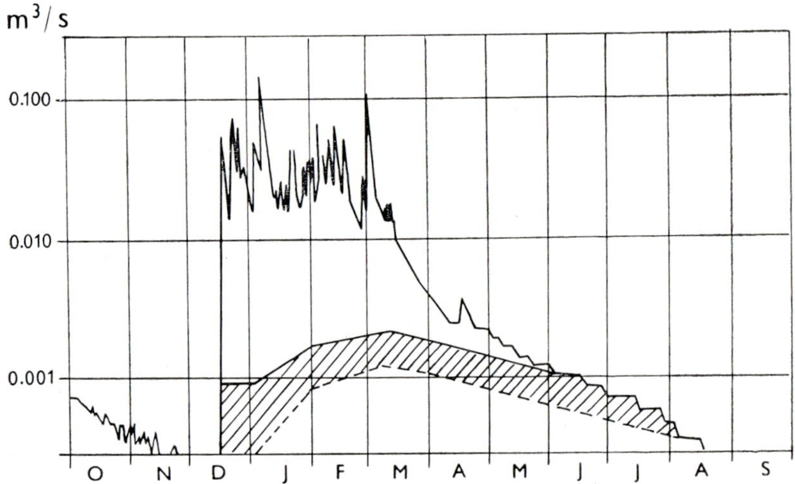

Fig. 6.4 Hydrograph of the outflow from an intermittent swamp indicating the prolonged effect of the surface runoff and the period of intermittency. The outflow from the groundwater storage in the swamp is hatched. The end of the rainy season was early in April

Vegetation plays an important role in the hydrology of swamps. It produces permanent changes in the depth and the direction of the channels inside the swamps and in the formation and floating of the islands. Taylor and Stewart [15] differed in their ecological approach to the permanent and seasonal swamp pattern. According to them perennial swamps are low-lying areas drained by large rivers. In the swamps of Papua and New Guinea they found as distinctive feature of riverside swamps the occurrence of various herbaceous communities at a distance of about 1 kilometre from the river. In seasonal swamps the same herbaceous communities were found only in the surroundings of pools of water. Kostermans, who studied peat swamps and their vegetation in the equatorial parts of Borneo, found the swamps in this region intersected by small areas of high land with distinctive types of vegetation reflecting the conditions of acid waterlogged sandy soils. In the more developed stage this was continuously replaced by true peat forest.

In Africa a difference was found between true swamps and the seasonal swamps known as dambos. While the dominant plants in African perennial swamps are *Cypherus Papyrus, Phragmites, Herminiear, Vossia suspidata, Pistia Stratiotes and Typha,* in intermittent swamps the following were found to be typical: *Aristida Atroviolaces, Brachyraria filifolia, Eragrostis capneisis, Eriochrisis purpurata, Hyparrhenia braceata, Hypogynium virgatum, Schizachyrium jeffroysi and Loudetia simplex.*

Verboom [16] found fourteen different species in the dambos of Zambia and Fanshave [8] found more than sixty species in just four dambos.

In coastal Asian swamps intensive rainfall contributes to the growth of species of mangrove. According to various sources of information the extent of the mangrove swamps is related to the erosional activity of the rivers and to the amount of the deposited mud. In this region the density of the mangrove population is clearly related there to the intensity of silt accumulation and to its type. Deposits of granitic origin result in a dense population, while those of sandstone origin are not too favourable for mangroves. Volcanic ash is a significant source of nutrients and component of swamps in various parts of Indonesia.

The same amount of rainfall distributed differently can contribute to the formation of different types of swamps. When conditions are unfavourable for bacteria's activity a peat forest unit can be formed.

A symbiosis of lakes and swamps is found in many parts of the tropics. The ratio of lakes and marshy regions can fluctuate year by year and season by season. Long term changes are also common. For instance, in Lake Valencia, Venezuela a century ago there was enough water to form a permanent outlet. Since then the water level has dropped by more than five metres and the lake has became bordered by swamps and marshy plants. There are shallow swampy areas in the middle of Lago de Izabel in Venezuela. Swampy belts along the African lakes form a very complex system and it is difficult to calculate the water balance of the lakes because a significant part of the inflow evaporates from the swampy areas before it can reach the lake. The combined effect of the Bangweulu Lake and swamps changes the regime of the River Chambeshi from that of an inflow into an outflow of the River Luapula. The water balance of these swamps remains active which means that some of the water from the swamp basin contributes to the total runoff, while in the swamps on the White Nile, a substantial amount of water inflow coming from the upper reaches is considerably reduced at the outflow.

The main role of the vegetation lies in the effect of its transpiration. Estimation of the partial effect of a single species can be studied indirectly by a comparison of the inflow and outflow. Such a type of analysis was made by Shahin [14] for the complex of the White Nile swamps, called Sudd Area, which includes part of Bahr el Gabal, Bahr el Gazal and Marchar marches (Fig. 6.5). The graph clearly

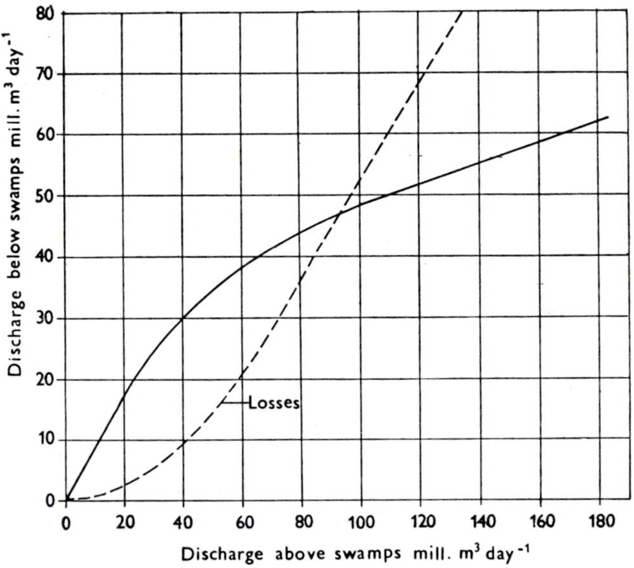

Fig. 6.5 Relationship between inflow into the Sudd Area, Nile basin and total water loss in the swamp, according to Shahin

indicates the water loss as being dependent on the inflow into the swamps. Here the varying access of aquatic plants to the water surplus possibly plays a dominant role. Hurst studied the role of the Nile papyrus and for the first time concluded that this plant can evaporate more than does the free water surface. Consumption also varies according to the type of plant and thus the actual water loss varies significantly in accordance with the prevailing type of vegetation and the proportion of free water and vegetation. Nevertheless, the inflow regime and climatic conditions remain equally significant factors.

Tab. 6.7 Water balance calculation for some African swamps

Parameter	Unit	Bangweulu swamp	Kafue Flats	Lukanga swamp	Headwater swamp
Drainage area	km²	102 000	58 290	19 490	1.43
Area of swamp	km²	15 875	2600	2600	0.15
Rainfall on dr. area	mm	1190	1090	1250	1330
Rainfall on swamp	mm	1210	1110	970	1330
Evaporation from free water surface	(mm)	2340	2070	2070	1710
Evapotranspiration outside swamps	(mm)	890	785	908	1320
Additionally evapotranspired from swamps	(mm)	1120—1260	196	252	—
Total evapotranspiration in swamps	(mm)	2000—2180	1000	1120	1075
Water loss in % of inflow		60	4	7.8	—

Tab. 6.7 shows the water balance for four different types of swamp in Africa. Bangweulu swamps are formed by standing or slowly flowing water, deep swamp vegetation, papyrus and Matete reeds and numerous small grassy islands. The Kafue swamps represent a lowland type situated along the banks of a great river. Such a type is saturated once a year by flooded rivers and dries out partly during the dry period. Lukanga swamp is in some ways a sidestream reservoir which has an identical inflow/outflow system. The excess water is stored there and flows out during the dry season. Dambo is a typical headwater swamp without any surface inflow. Rainwater is temporarily stored and slowly released during the dry season.

There is no doubt that different regime can be found for morphologically similar swamps, under different climatic and hydrologic conditions. Thus to some extent the vegetation can be considered to be a secondary product of more complex conditions.

Another type of information can be derived from the monthly fluctuation of evaporation from swamps. Tab. 6.8 shows the monthly values of the evapotranspiration from Bangweulu swamp and from the headwater intermittent swamp located nearby. The evapotranspiration from the headwater swamp is much lower than the evapotranspiration from the permanent swamp, although the total rainfall is almost the same in both swamps.

In general, water consumption by the vegetation is much higher in the perennial swamp during both the dry and the wet periods, although during the latter, watered areas are found in both types of swamps. During the dry period, the watered area

Tab. 6.8 Comparison of monthly evapotranspiration from neighbouring perennial and intermittent swamps

Month	Swamp evapotranspiration (mm)	
	Perennial	Intermittent
10	114	28
11	261	90
12	249	94
1	250	97
2	261	69
3	276	53
4	208	26
5	122	10
6	97	9
7	101	6
8	107	5
9	109	5
Year	2156	494

in the intermittent swamp gradually decreases and the transpiration rate is greatly reduced.

It can be concluded that the presence of swamps in the basin greatly reduces the total runoff. Smoothed values of the runoff depending on the percentage of swamps in the basin and on the annual rainfall are given in Fig. 6.6. Deviations can be expected from case to case for the same percentage of swamps, this being dependent on other factors.

Modern methods have been introduced into the analytical work concerned with the hydrological regime of swamps. Dincer et al. [7] studied the regime of the Okavango swamp in Botswana using isotope techniques. The swamp is located in the region where the river Okavango deposits all its sediment load and 95% of the water is lost through evapotranspiration. The analysis of stable isotopes proved that the central distributory system is more active than the swamp peripheral and that there is no significant groundwater outflow which indicates that almost all the loss is due to evapotranspiration. In winter when a high water level occurs, loss from swamps is almost entirely due to evaporation, while in summer when the water levels are low, evaporation and transpiration make an almost equal contribution. The main balance components obtained through isotope analysis are shown in Tab. 6.9.

Tab. 6.9 Water balance of Okawango swamp, according to Dincer, Hutton and Kupee [7]

Mean area of swamp	10 000 km^2
Max. area of swamp	13 000 km^2
Min. area of swamp	6000 km^2
Mean active storage	4×10^9 m^3
Max. active storage	7×10^9 m^3
Min. active storage	1×10^9 m^3
Inflow	10.5×10^9 m^3
Precipitation	5×10^9 m^3
Evapotranspiration	14.9×10^9 m^3
Outflow-surface	0.3×10^9 m^3
Outflow-groundwater	0.3×10^9 m^3

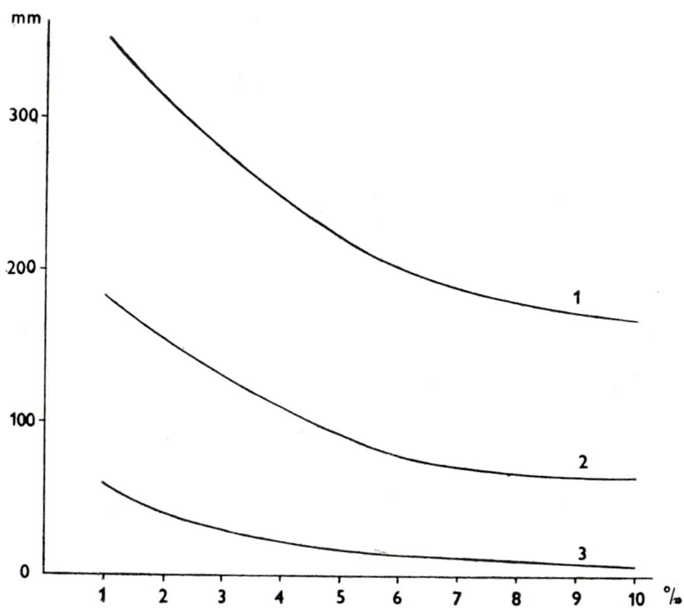

Fig. 6.6 Relationship between total runoff from the basin, percentage of swamps in the basin and mean annual rainfall. 1, annual rainfall 1250 mm; 2, 1000 mm; 3, 750 mm

One of the first attempts to simulate the water regime of a large African swamp was made by Hutchinson and Midgley [10]. A complex of the river and swamp system in the lower reaches of Okavango, Botswana, was simulated by the Muskingam method. Severe earth tremors in 1952 produced changes in the gradient of the swamps which made simulation rather difficult. The authors felt that ecological studies should be undertaken in swamps before further progress in simulation could be achieved.

A headwater swamp regime was simulated by a deterministic approach (Balek [1]). The main problem was to simulate the formation of the intermittently waterlogged soil profile and to separate the surface runoff from the baseflow.

Sellars [13] developed a model for the simulation of evaporation losses from the flood plains of the upper Yobe River, northern Nigeria. He found that 70% of the total annual runoff is lost by evaporation during the flood season and the effect lasts until the dry season because of the storage effect on floodwater. The swamp system simulation was based on the relationship between the storage and the flooded area. The following continuity equation was derived:

$$ds(t)/dt = I(t) - Q(t) - R(t) - E(t)\,A(t) - V(t)[A_{max} - A(t)] + \\ + P(t)\,[FH + A(t)\,(1 - F)].$$

where $S(t)$ is for storage, $I(t)$ for river inflow, $Q(t)$ for river outflow, $R(t)$ for groundwater losses, $E(t)$ open water and swamp evaporation, $V(t)$ evapotranspiration from crops, $A(t)$ area flooded, A_{max} maximum area flooded, $P(t)$ precipitation, F runoff coefficient, H total area of flood plain.

The preparation of adequate data at sufficiently short intervals appears to be the main problem of such an approach. At present many problems relevant to swamp hydrology remain unsolved. They can be listed as follows:

a) evaluation of total water resources and determination of possible rates of water consumption in the complex utilisation of water resources;

b) evaluation of the role of swamps as a natural factor in recharging rivers and in the water balance impact;

c) estimation of water consumption and root development in accordance with the swamp development;

d) water balance as relevant to topography, structure and vegetation;

e) optimization of water resources utilisation in swamps and their environment.

6.4 REFERENCES

[1] Balek, J., 1975. A year round water balance model, its application and its future. Int. Symp. on Math. Models in Hydrology. IAHS Publ. No. 115, 6—12.

[2] Balek, J., Perry, J., 1973. Hydrology of African headwater swamps. Journal of Hydrology 19, 227—249.

[3] Balek, J., 1977. Hydrology and water resources in tropical Africa. Elsevier Amsterdam, 208 p.

[4] Barabas, S., 1981. Eutrophication can be controlled. Water Qual. Bull., Vol. 6, No. 4, 94—155.

[5] Cullen, P., Smalls, I., 1981. Eutrophication in semi-arid areas. Water Qual. Bull., Vol. 6, No. 3.

[6] Debenham, F., 1952. Study of an African swamp. Colon. Off. London, 52 p.

[7] Dincer, T., Hutton, L. G., Kupee, B. B. J., 1978. Study, using stable isotopes of flow distribution, surface—groundwater relations and evapotranspiration in the Okavango Swamp, Botswana. Isotope Hydrology. IAEA, Vienna, 3—26.

[8] Fanshave, D., 1971. The vegetation of Luano catchments. NCSR WR 10 Rep., Lusaka, 36—47.

[9] Hurst, H. E., 1954. Le Nil. Payot, Paris.

[10] Hutchinson, I. P. G., Midgley, D. C., 1973. A math. model to aid management of outflow from the Okavango swamp. Journal of Hydrology 19, 93—112.

[11] Kalef, A. T., 1981. The Nile—one river and nine countries. Journal of Hydrology, Vol. 53, 53—72.

[12] Kimble, G. T., 1960. Tropical Africa. The 20th Cent. Fund, New York.

[13] Sellars, C. D., 1981. A floodplain storage model used to determine evaporation losses in the Upper Yobe River, northern Nigeria. Journal of Hydrology 52, 257—268.

[14] Shahin, M., 1971. Hydrology of the Nile basin. Int. Courses in Hydraulics and San. Eng., Delft, 138 p.

[15] Taylor, B. W., Stewart, G., A., 1958. Vegetational mapping in the territories of Papua and New Guinea, conducted by the CSIRO. Proc. of Kandy Symp. on Trop. Veget., Unesco, Paris, 127—136.

[16] Verboom, W. C., 1965. Aerial photographs, relics of soil, vegetation and tse-tse flies, I.T.C. Delft, 21 p.

[17] Welch, I. J., 1952. Limnology. McGraw Hill, New York.

[18 Anon., 1976. World water balance and water resources of the earth. (In Russian). Gidrometeoizdat, Leningrad, 630 p.

7 HYDROLOGICAL EXTREMES

7.1 GENERAL

The occurence of hydrological extremes in the tropics is evaluated differently from those occurring under moderate conditions. Floods which are usually considered to be disasters are taken as a bless'ng under semi-arid conditions and are awaited long before the time of their arrival. In some regions the fertility of the soil depends on the magnitude of seasonal floods. For instance the Red River which is more than 1200 kilometres long and rises in China, meanders in its last 220 kilometres through a deltaic flood plain which is sown almost entirely in the form of paddies. The delta lies below the flood level and only that part of the flooding which is in excess of paddy consumption is fought against.

Rural water tanks in arid regions of wet and dry climate are also anticipated and agricultural practices and crops, herds and game are adapted to the fluctuation of the wet and dry seasons. Nevertheless, there are some years when the dry period is too prolonged or there is not enough rain between two dry seasons conversely, there are other years when exceptional floods cover extensive areas with a sheet of water.

Many attempts have been made to define extremes, or at least the boundaries between extremes and a situation which can be considered normal. In simple schemes a difference is made between so-called high water, low water and normal water years. Starmans and Shalash [29] defined a high water year as a year in which the mean annual river stage or mean annual runoff is higher than in a normal year which, however, does not mean that the river stage or discharge throughout the whole high water year should always be above normal.

Similarly, a year which has a mean annual river stage or discharge lower than the normal year is regarded as a low water year. Again, the water levels or the discharge during considerable periods of a low year can be higher than the corresponding values in a normal year.

A highest water year is a year in which the highest flow or stage is recorded and this year may not coincide with the high water year. The lowest water year is the year which records the lowest flow or lowest water level.

From simple verbal description the development of the analysis of the extremes has become more and more sophisticated and a considerable amount of analytical work has been accomplished and sophisticated formulae drawn up. However, only long-term observation of the extremes can form a solid base for such a work.

Unfortunately, only few rivers in the tropics have been observed over a period of at least fifty years and there are many gaps in the records. Thus it is often difficult to analyse properly the sequence of extremes for the needs of the constructional engineer. In some cases a hydrologist in the field finds himself dependent on oral records, which are not always very reliable, although they still provide some basis for sound estimation.

7.2 FLOODS

Flood analysis is closely related to consideration of risk and reliability in hydrological design. Cordery [8] specified two considerations as being the most fundamental:

a) What are the loadings on the structure?

b) What degree of safety should be incorporated in the design?

From the hydrological point of view the loading means the flood which can pass through the structure. If flood estimate in the design is incorrect then the benefits of subsequent hydraulic and structural design are largely nullified. It is typical for flood loadings that increased safety can be obtained by increasing the capacity of the structure at a considerably increased cost. This holds good for small as well as large structures. In the tropics major studies are usually undertaken for estimating spillway design floods for important dams, but very little effort is made to estimate properly design floods for small structures. However, as has been found in Australia, more than 60% of all expenditure is on water works which taken individually are of minor importance. Therefore the design frequency analysis is a most important step in the design of flood passing structures. Only very large and important structures can be designed on the basis of probable maximum, but the majority are designed to pass something less than probable maximum flood, for economic reasons.

The selection of design frequency depends upon many factors. Design based on a flood of selected probability implies that some damage will occur at every occurrence of a flood of lower probability and it is a matter for economic consideration what sort of damage can be permitted for the construction and how often.

The design of the formulae for the calculation of flood peaks of certain probability can be traced back to the middle of the nineteenth century (Chow [7]). The main difficulty with the application elsewhere of various types of empirical formulae is the proper selection of the parameters involved. The extension of the validity of the formulae to tropical conditions is even more difficult, because of their derivation in hydrological and meteorological conditions very different from those in the tropics. However, many types of formulae and methods for the estimation of flood peaks of various probability developed in moderate regions have been applied in tropical countries.

For instance, in India three methods of peak rate flow calculation have been favoured (Gupta et al. [13]), namely the rational method, Cook's method and the so called 'hydrological soil cover complex number' method (see Ogrosky, Mockus [22]). Cook's method (Calp [5]) was designed by the Bureau of Agricultural Engineering in the U.S.A. and conditions of aridity have been taken into consideration in the calculation. The rational formula is used in the form

$$q = 0.28CiA \quad \mathrm{m^3\,s^{-1}}$$

where C is runoff coefficient, i is rainfall intensity (mm $\mathrm{h^{-1}}$), of same frequency as the flood discharge, or duration equal to the time of concentration and A is the watershed area ($\mathrm{km^2}$).

Many tables are available for an estimate of C and a wide variety of C can be found for the same watershed. Rainfall intensities are more frequently available from the Meteorological Offices, however, several authors have concluded that the frequency of rainfall should not be considered as the same as flood frequency. Sometimes the values of C and time of concentration are considered as relevant values.

Batista [3] applied the empirical formulae for the peak flow calculation developed by Gumbel, Nash and Lebedev on the rivers of Cuba. He compared the results with the empirical formula and as could be expected the results for the same river cross-sections differed. Unfortunately, in such a case it is not possible to conclude definitely which of the formulae can be recommended as most suitable for the tropics. In fact the selection of the formula depends mainly on the type of data immediately available.

Similarly Monteguado [21] adapted the flood peak calculation method of Alexeev to Cuban rivers. The method is based on a determination of the time of travel and requires the application of several tables, which can be found in the cited literature.

It would be impossible to present all formulae available with an explanation of them. This can be found in basic hydrological literature. The selection of the proper formula should be verified on some river in the studied region, for which a longer reliable record is available.

More valuable are the results of flood studies performed directly in tropical regions. Particularly useful are studies with conclusions on the regional validity of the results. Significant flood studies in west Africa have been carried out by French teams, particularly by ORSTOM. Several experimental basins in west Africa had already been established for this purpose before the International Hydrological Decade started and many results related to each particular basin were published.

Attempts were made from the beginning to establish the relationship between effective rainfall and the formation of hydrographs. In the Sahel region Rodier [24] studied the basic characteristics of effective rainfall producing flood hydrographs. A difficulty in carrying out such a study lies in the non-availability of autographic records; normally only daily totals are available. Therefore, in accordance with the estimation of permeability, the limits of infiltration conditions were established which when combined with the daily totals, gave basic information on the rainfall – runoff relationship. For impermeable soils an infiltration value of 10 mm $hour^{-1}$ and for permeable soils a value of 40 mm $hour^{-1}$ were estimated; in regions with an annual rainfall of 300 – 1000 mm, it was calculated that effective rainfall, up to 85% of the daily total, can fall in 90 minutes on soils accepting 20 mm $hour^{-1}$ and 75% of the daily total can fall in 55 minutes on soils accepting 40 mm $hour^{-1}$. Providing the rainfall record is long enough, such a study can serve as a basis for the calculation of effective rainfall with a certain probability of occurrence.

In the initial studies carried out in southwest Africa, the flood regime was studied with regard to the number of years in which the river flow reaches the ocean. The River Swakop, as described by Wipplinger [31], flooded once in six years, as did the River Gamams. According to the author, a certain characteristic of the flood regime is the marked degree to which flood waters are dissipated on their path down into the sandy channels. The flow of the river into the sea occurs very seldom owing to the dissipation. This is, of course, a very special approach to flood frequency analysis, influenced by extremely arid conditions and by the ephemeral regimes of the wadi type of river. However, in arid regions, a similar analysis can be undertaken not only for the mouths of rivers, but for any other cross-section. Beside the occurrence of the event, its length can also be used as an additional characteristic, when the actual amount of water in sandy river beds is unknown. Thus the season of 1934/35 was considered as particularly important for the Swakop River, because the river then flowed continuously for four months.

The empirical probability curve is often used for the calculation of flood frequency:

$$p = \frac{m}{n+1} \qquad 100\%$$

where m is the number of events in the descending sequence of annual flood peaks and n is the total number of events. The formula is valid in the case of one river flood occurring once a year. A simple approximate formula has been developed for analysing the flood regime of tropical rivers reflecting the possibility of the occurrence of more floods in one year, or in other words events not equalling the number of years (Balek, Holeček [2]). This is particularly useful for rivers with a double flood regime and also for ephemeral streams. Tab. 7.1 gives a sequence of seventeen culminations of the River Pra, Ghana. Here one maximum per year was taken into account and the probability curve was calculated. In Tab. 7.2 the sequence of all floods observed within the same period was analysed and the theoretical curve plotted in Fig. 7.1. The probability scale was re-evaluated by using the formula

$$p = \frac{100}{N - \frac{n}{M}} \quad \%$$

where p is the probability of the occurrence of the flood being repeated once in N years, n is the number of events and M is the number of years of observation.

Tab. 7.1 List of floods on River Pra, Ghana, 1944—1960, one flood per year included, empirical probability calculated

No.	Month	Year	Q_{max} $m^3\,s^{-1}$	$p = \frac{m}{n+1} \cdot 100\,\%$*)
1	7	60	1280	5.57
2	7	57	1020	11.14
3	7	53	990	16.70
4	9	47	810	22.30
5	11	55	780	27.85
6	6	58	780	33.40
7	6	56	744	39.00
8	10	51	720	44.50
9	7	44	660	50.10
10	7	49	660	55.70
11	9	52	570	61.20
12	10	59	550	66.80
13	10	45	504	72.40
14	6	48	504	78.00
15	7	54	504	83.50
16	10	46	420	89.00
17	10	50	262	94.60

*) m = number of event
n = total number of events

The results for the flood regimes of the humid tropics with two maxima in a year and a simulated ephemeral stream with six maxima in 17 years are given in Tab. 7.3 and plotted in Fig. 7.1. Here we can see how the results differ for the highest and lowest probabilities.

Pearson's type III distribution was used in the described analysis, this having been found most convenient for flood frequency analysis in the temperate regions. McMahon and Srikanthan [19] endeavoured to ascertain whether such a type of

Tab. 7.2 List of floods on River Pra, Ghana, 1944—1960, all floods included

No.	Month	Year	Q_{max}	$p = \frac{m}{n+1} 100\,\%$
1	7	60	1280	2.86
2	7	57	1020	5.70
3	7	33	990	8.60
4	10	60	840	11.40
5	9	47	810	14.30
6	11	55	780	17.10
7	6	58	780	20.00
8	6	56	744	22.80
9	10	51	720	25.70
10	7	55	572	28.60
11	7	44	660	31.50
12	7	49	660	34.30
13	10	49	635	37.20
14	10	53	605	40.00
15	9	52	570	42.80
16	6	52	550	45.70
17	10	57	550	48.50
18	10	59	550	51.40
19	5	59	505	54.40
20	10	45	504	57.20
21	6	48	504	60.00
22	7	54	504	62.80
23	11	54	460	65.20
24	11	55	436	68.60
25	5	59	436	71.40
26	10	46	420	74.20
27	10	44	380	77.20
28	10	56	366	80.00
29	11	48	282	82.80
30	10	50	262	85.60
31	7	45	250	88.70
32	6	50	250	91.40
33	7	46	232	94.20
34	10	58	196	97.20

distribution is suitable for arid and semi-arid conditions. When comparing this distribution with other theoretical distributions, they found it most convenient for a variety of Australian rivers, while log-normal, gamma, Gumbel's and Weinbull's and exponential distribution were inappropriate for the data.

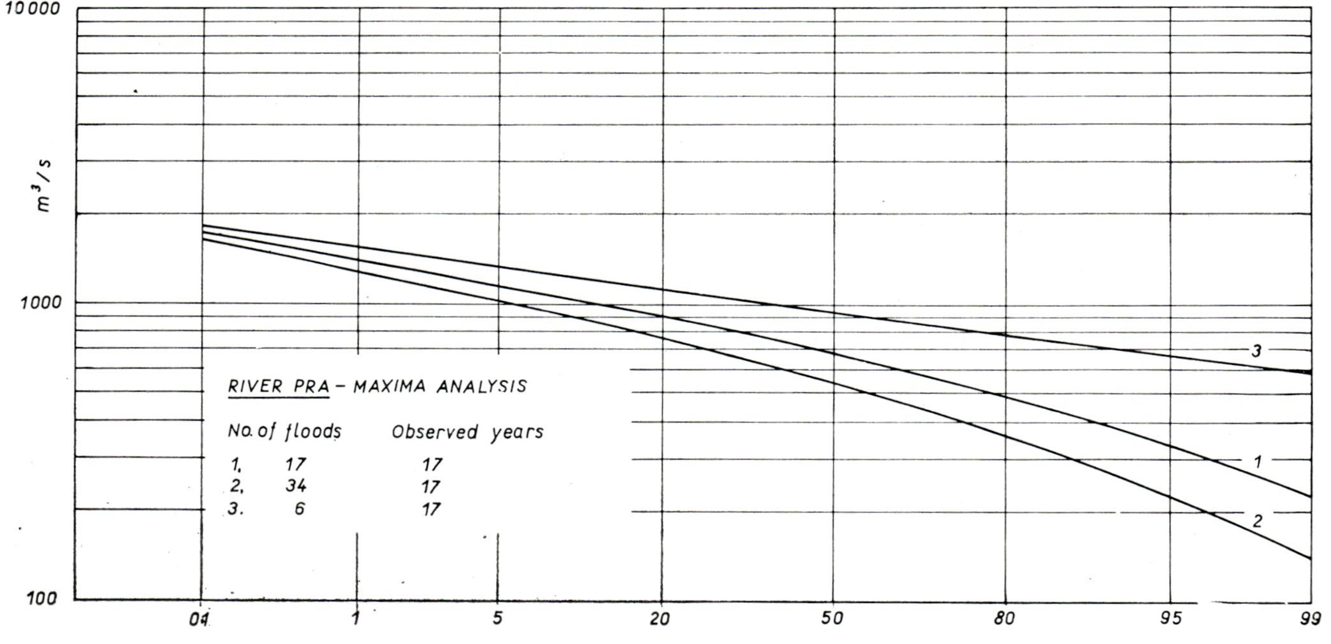

Fig. 7.1 Perason's type III curve for the calculation of discharge maxima of the River Pra. 1, 17 years of observation and 17 events; 2, 17 years of observation and 34 events; 3, 17 years of observation and 6 events

Tab. 7.3 Floods as calculated for various probabilities and number of events, River, Pra, Ghana, 1944—1960

1 flood in a year		2 floods in a year		6 floods in 17 years	
N years	*Q* $m^3 s^{-1}$	*N* years	*Q* $m^3 s^{-1}$	*N* years	*Q* $m^3 s^{-1}$
100	1710	100	1670	100	1700
50	1550	50	1510	50	1580
20	1320	20	1290	20	1480
10	1020	10	900	10	1300
5	900	5	780	5	1050
2	680	2	540	2	950

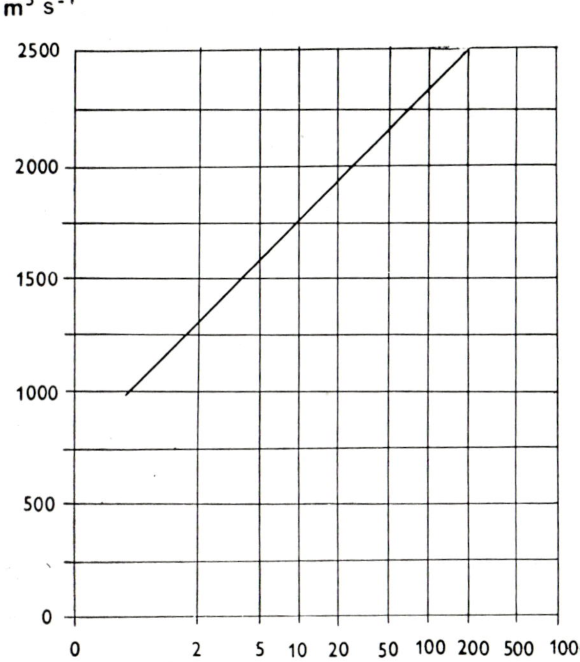

Fig. 7.2 Frequency analysis of the annual floods for the river Quevedeo, Ecuador, according to Molina

For a similar purpose, Molina [20] analysed the flood frequency of the Ecuadorian rivers. He found the empirical curve as plotted in Fig. 7.2 most fitting.

Because not even short sequences are yet available for many African streams, regional formulae have been developed for some regions. Pitman and Midgley [23] applied Hazen's method and the Myer–Jarvis formula to the conditions found in South Africa and divided the country into seven regions:

1. High rainfall areas, mountains, a high proportion of exposed rocks, sclerophyllous vegetation.

2. Interior plateau, relatively high rainfall, early summer thunderstorms, grassland, fairly dense drainage pattern, parts suffering from advanced soil erosion.

3. Year round rainfall, dense drainage pattern, high proportion of exposed rocks, advanced soil erosion.

4. Eastern coastal strip, high rainfall, dense drainage pattern, generally well vegetated.

5. Subhumid to semi-arid interior plateau, rains generally late summer.

6. Well-forested bushveld areas, low rainfall, areas well forested.

7. Arid interior and west coast, low rainfall, areas well forested.

Tab. 7.4 Flood peak/mean annual flood peak ratio for various recurrence intervals and South African regions

Region	Flood peak/mean annual flood peak for recurrence intervals (years) of				
	5	10	20	50	100
1	1.54	2.08	2.67	3.54	4.28
2	1.70	2.43	3.25	4.48	5.54
3	2.00	3.14	4.51	6.68	8.60
4	1.97	3.02	4.24	6.13	7.78
5/7	2.04	3.22	4.58	6.82	8.78
6	2.46	4.40	6.82	10.75	14.26

Although only zone 6 is located in the tropics, the results as supplied in Tab. 7.4 can be extended north of the Tropic of Capricorn, provided the similarity of the regions and regimes has been proved and the mean annual flood peak regime is known.

The mean annual flood, as related to the catchment area was analyzed by Kovács [14] for some east African catchments. The author compared the water yield of mean annual floods with the catchment area and differentiated between the catchments of limited area, that is, catchments influenced by heavy rainstorms of high intensity on small areas, and catchments of larger size under the influence of less intensive rainfall. As indicated by Fig. 7.3 the difference in length of the horizontal part of the curve is the main characteristic of each particular regime.

A flood analysis of Asian and African rivers was made by Starmans [28] who developed the graphical relationship between the drainage area and so-called peak discharge (Fig. 7.4), a term related to a 1–2% flood.

A comparison between the 1% flood as calculated for some African basins and South African and European 1% floods was derived by Balek [1] from various sources. The results are found in Fig. 7.5 and the validity of the curves is described in Tab. 7.5. Such an approach usually leads on a regional scale to the development of the regional formula type of

$$q = \frac{b}{A^c}$$

where q is the 1% flood peak yield ($1\ s^{-1}\ km^{-2}$) and A is the drainage area, km^2 and b and c are coefficients characterizing the basin.

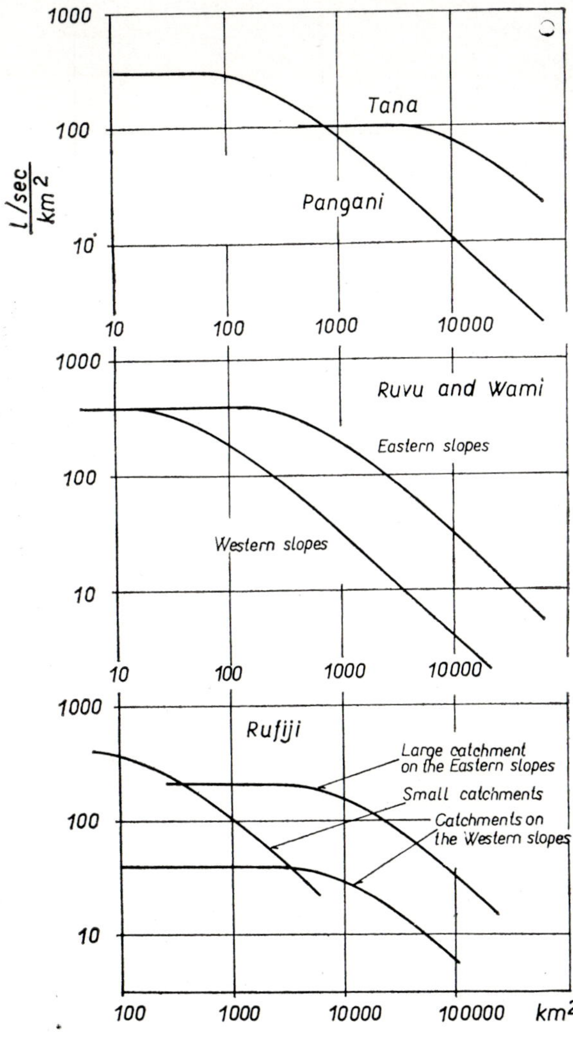

Fig. 7.3 Relationship between mean annual flood and drainage area for East African rivers, according to Kovács

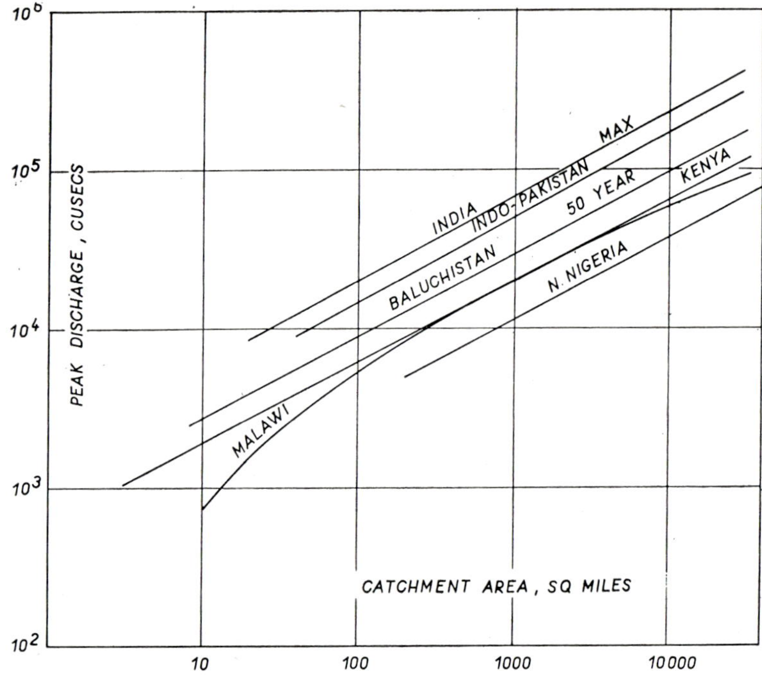

Fig. 7.4 Relationship between drainage area and 1—2 % probability of occurrence of peak discharge for African and Asian rivers, according to Starmans

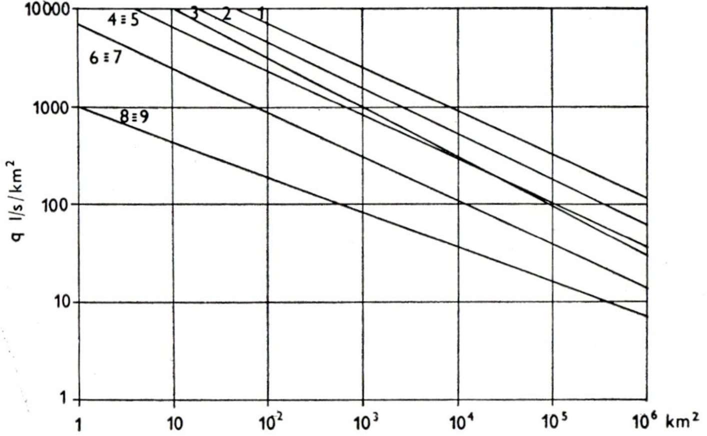

Fig. 7.5 Relationship between the drainage area and the yield of 1 % flood, based on observation of African and European streams

Tab. 7.5 Tropical and European regions having different flood regimes

1. Insufficiently afforested and denuded areas as in the mountainous parts of the Mediterranean.
2. Mountainous areas and deep valleys as on the Central African Plateau.
3. Mountainous regions as in South Africa.
4. Mountainous regions as in Central Europe.
5. Streams with pronounced rolling character of the basin.
6. Flat coastal tropical areas.
7. Tropical rivers with flat or less pronounced rolling character of the basin, not influenced by swamps.
8. Flat basins as in Central Europe.
9. Flat tropical basins influenced by swamps.

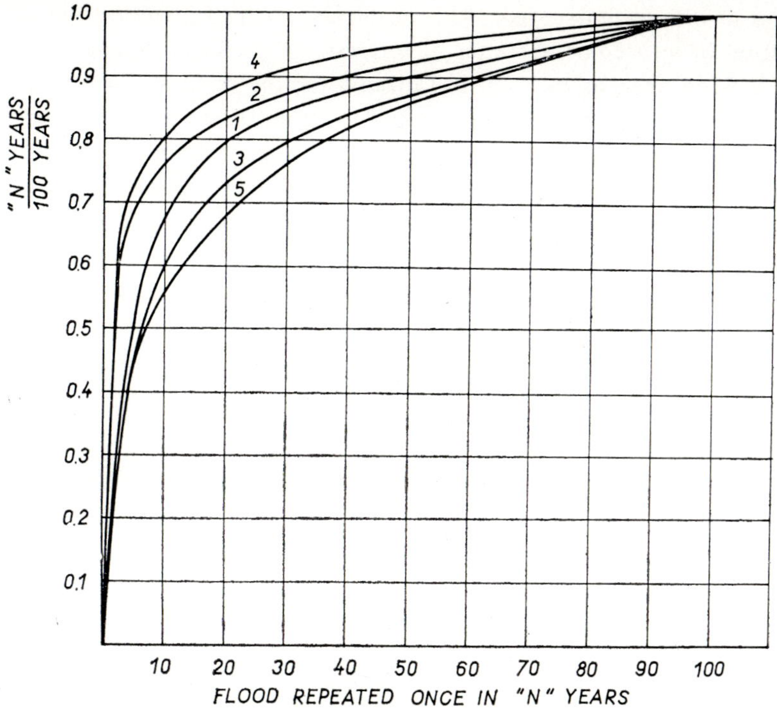

Fig. 7.6 Repetition of floods as related to the flood repeated once in 100 years

Fig. 7.6 serves as a calculation of the recurrence interval of various floods as related to a flood repeated once in 100 years. The curves are typical for the regions as described in Tab. 7.6.

An incredibly high flood peak yield for the lowlands of 23 $m^3\ s^{-1}\ km^{-2}$ was observed in Sierra Leone by Ledger [17]. This would be more typical for mountai-

Tab. 7.6 Regions for which the characteristic curves are shown in Fig. 7.6

1. Streams with pronounced rolling character of the basin not influenced by swamps.
2. Rolling basins of tropical plateaux.
3. Mountainous basins and escarpments.
4. Undulating basins.
5. Flat basins.

nous areas; at least it indicates how carefully any regional analysis has to be applied outside its own boundaries.

Still, in many developing countries designers of water constructions have very little data for the design, however, there is very little capital to allow construction based on an unacceptable risk. The problems of the full utilisation of scanty information have become a matter of interest. Schiller and Ngana [25] tried to develop a flood analysis method in Tanzania for the regions where hydro-power schemes are to be constructed. For that purpose they tried the so-called TRRL east African flood model (Greening [12]), which is a reservoir analogue model consisting of two parts. The first part simulates the process between the rainfall and runoff entering the system, the second is a finite difference routing technique which simulates the passage of the flood wave through the channel system. The model serves for the calculation of the 10 years flood and Schiller and Ngana developed a regional relationship between the ratio of the peak value and so-called index or mean annual flood (Fig. 7.7) and between the drainage area and the mean annual flood (Fig. 7.8). By using both curves and knowing a 10 years flood, one can extend

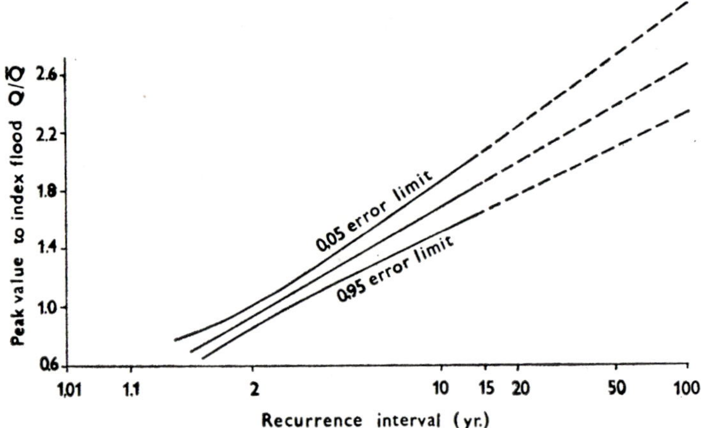

Fig. 7.7 Regional flood frequency curve for Mbeya region, Tanzania, according to Schiller and Ngana

the prediction of floods to lower probabilities of their occurrence. Both graphs were developed for the Mbeya region in Tanzania. The successful application of the above method for another region depends on the ability to derive a curve analogous to the one in Fig. 7.7.

Often the occurrence of floods is not solely responsible for damage to water constructions. Dake [9] found that due to considerable erosion in the river bed at the outlets of the spillways and other structures during some minor floods, substantial damage to the spillways and other structures can occur. Thus without any records of the lower peak flow the probability of its occurrence can be overestimated.

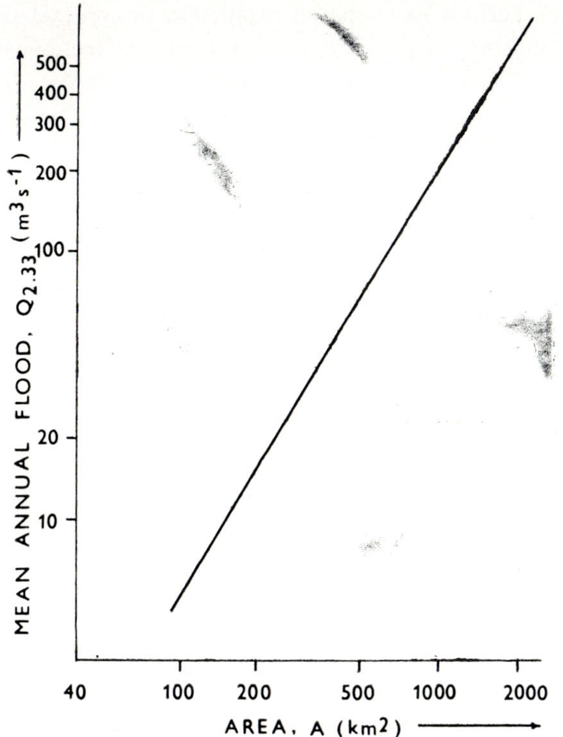

Fig. 7.8 Relation between area and mean annual flood for Mbeya region, Tanzania, according to Schiller and Ngana

7.3 DROUGHTS

Droughts in the tropics can produce more disastrous situations than floods. Unfortunately, from the hydrological point of view, the effects of drought are far less known and statistical analysis results fewer than for floods and prevention is much more difficult. The most significant difference probably lies in the duration of the extremes. While the flood peak is a matter of several hours, perhaps days, the maximum effects of droughts may continue for years. Under a moderate climate and with perennial rivers, droughts can be defined by the lowest stage of a river at a certain cross-section. Under arid climates, however, there are many rivers which regularly dry out even in the 'normal' years. The duration of the intermittent period then becomes the most significant characteristic. It is natural that many tropical rivers dry out regularly for several months and only a prolonged period is considered as catastrophic, while the catchments are ecologically well adapted for the normal dry period.

The intensity of droughts is judged not only by the occurrence of low flows in the rivers but the impact on natural vegetation, crops, livestock and game. Here the annual rainfall and its seasonal distribution is of primary importance. According to Gibbs and Maher [11] rainless periods have spatial expression in contrast to floods and river flow and soil moisture regime becomes a better hydrological indicator of the impact of drought. Under arid conditions soil moisture replenishment is very temporary and thus plant growth, too is intermittent as in Central Australia. The dry periods are long and unpredictable. The periods when soil moisture is favourable for plant growth are few. At Alice Springs, Australia, 100 days of growth occur in a calendar year. Slatyer [27] estimated that in the Australian seasonless zone the rainfall necessary to start plant growth occurs three times a year and only two of these occasions are followed by sufficient rain to support continued growth. Only woody plants replace their leaves and the seed of herbaceous plants can germinate and retain enough water for further survival.

Riparian vegetation in the arid regions extracts a large amount of sensible heat from air moving from the adjacent slopes so that it may transpire as much as 1500 mm per year. Simple clearing is soon replaced by new vegetation and this is regarded as another feature of the oasis effect. In such a type of stand the so-called clothesline effect is also observed, when irrigated plants are widely spaced and thus a greater heat transfer of hot air can occur.

Whenever drought is discussed, the problem of its definition arises. Extreme drought is sometimes defined as a state of low annual rainfall. Other sources include in the definition the seasonality of the dry periods. Another approach defines drought in terms of normal rainfall and thus considers the occurrence of drought to be when the rainfall is less than an arbitrarily chosen proportion of the long term average. Yevjevich [32] stated that drought is a meteorological phenomenon that occurs when precipitation and/or natural runoff in a given period are less than normal and when this deficiency is great and prolonged enough to damage human activities.

Cervera and Aries [6] analysed the occurrence of droughts in Mexico and listed typical features relevant to drought analysis:

a) manifestations of the phenomenon are not easily seen and can only be predicted and appreciated through their consequences.

b) phenomena typical of drought have different significance for different specialist, thus preventing the establishment of uniform criteria that would explain the dynamics of the processes.

For a severe drought in Mexico in the period 1977 – 1978 the following characteristics were found:

a) low storage level in the reservoirs whose collecting basins are in rain-poor areas;

b) serious effects on the irrigation systems in double crop areas,

c) reduction of normally expected generated energy,

d) severe impacts on agriculture and livestock.

e) occurrence of forest fires.

Obviously economic aspects accompany any evaluation of droughts. An evaluation of Mexican droughts can be seen in Fig. 7.9 in the form of an area-intensity relationship established for the most severe droughts in forty years.

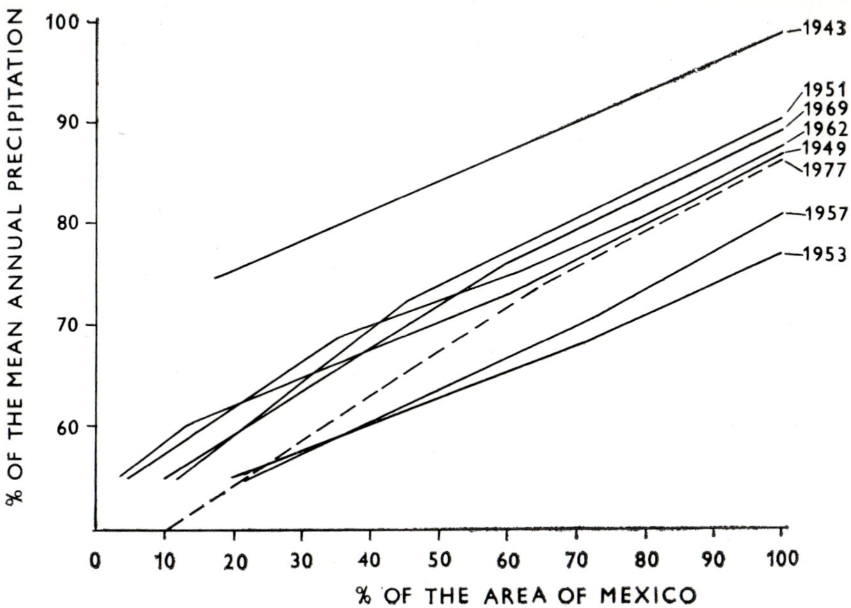

Fig. 7.9 Area-intensity relationship of the most severe droughts in Mexico, according to Cervera and Arias

The occurrence of droughts has several sources. A great part of Africa experiences the effects of continentality; along the Andes what is most decisive is the rainshadow, in Sahel an important role is played by the high pressure belt, in Chile and Namibia cold oceanic currents along the coast have a decisive effect and for conditions of extended aridity an important role is played by the history of soil development and land use. The last influence is mainly to be seen in the unsuitability of soils to store rainfall because of their overheating [16] and in the agricultural imbalance. The semi-arid areas contain half of the world's stock of cattle, 33% of sheep and 66% of goats, however, their productivity is only 20% of total world production. Another common feature is low annual precipitation and its frequent deviation from the mean annual rainfall which can easily exceed 100%.

Human activity in regions subject to the frequent occurrence of drought is adapted to a far smaller amount of rainfall and runoff than in the moderate regions, so that what is a drought in a moderate region is still a normal situation under semi-arid conditions. When, however, a period of large scale drought occurs, the situation becomes more critical than elsewhere. In addition to the well-known Sahelian drought in 1968 – 1974, in which more than a hundred thousand people died of starvation, a severe drought hit Central America due to deficient rainfall in 1977 and 1978.

Cervera and Arias [6] analysed the history of droughts in Mexico and found fifty documented cases of droughts in the Valley of Mexico and 26 cases in the Bajio area between 1521 and 1821. Seven of them lasted for two years and two for three consecutive years. Between 1875 and 1910 there were 29 prolonged droughts, in some regions such as Tlaxcala, Puebla, Hidalgo and Queretaro, the drought continued for seventeen years.

Many attempts have been made to forecast droughts, but, so far the application of various methods has not been completely successful. An analysis based on historical events was made by Krishnan [15], Tabony [30] and Beran [4]. It was concluded that there is some relationship between large-scale synoptic weather patterns and the occurence of drought and that there is another relationship between large – scale geophysical feature and drought. For instance the drought on the Indo-Gangaic plain is related to mid-troposphere divergence in the monsoon season, but the divergence cannot be sufficiently predicted in advance. Schereschewsky [26] suggested that in some regions sea surface temperature anomalies could be utilised for long-term weather prediction and the results could be further utilized for forecasting drought.

Soil water storage and soil water deficit have been used as indicators of so-called hydrological drought. Cordery [8] assumed that the simulation of soil water deficit as a result of special meteorological conditions could be utilised for drought analysis. However, among hydrologists drought analysis based on the river regime is most popular. McMahon [18] analysed the hydrological characteristics of arid zones and concluded that arid zone streams are considerably more variable than those in humid regions and therefore the hydrological characteristics based on humid region data should not be extrapolated to arid zones. He also found that maximum potential regulation for most of the arid zones is generally less than 10% of mean annual flow. The distribution of the annual flow of arid

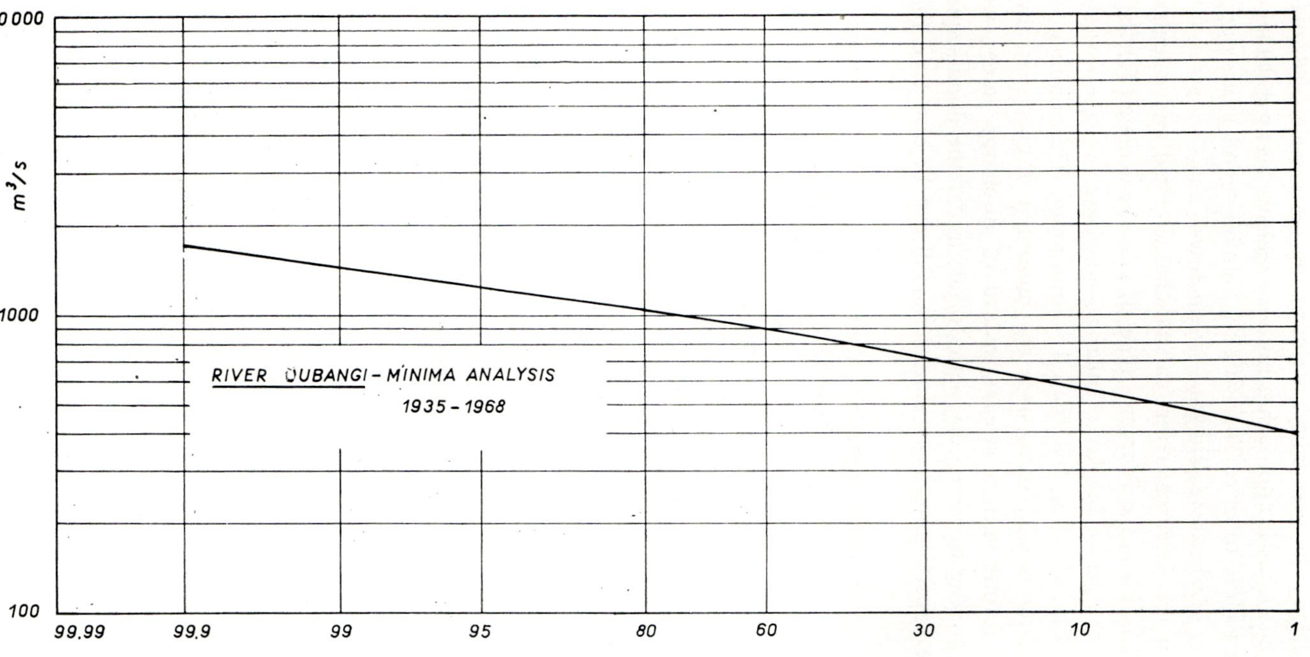

Fig. 7.10 Pearson's type III curve for the calculation of the minima of the River Oubangi, Central African Republic

zone streams tends to be more skewed than the flow distribution of humid zone streams. Considering the possibility of simulating the regime of drought, it was concluded that a stochastic approach is not very realistic because the lag one serial coefficient is very close to zero. Therefore the rainfall-runoff models appear to be more realistic for synthetising arid zone streamflows.

The regime of the perennial rivers can be characterized and calculated as for maxima. Fig. 7.10 shows the Pearson's type III curve, calculated for the sequence of minimum discharges of the River Oubangi, indicating that the lowest recorded discharge of 344 $m^3 s^{-1}$ was less than 1 % minimum. The minima analysis of intermittent streams is simplified, since the minimum discharge is always zero; when studying an intermittent stream, instead of discharge values, we analyse a sequence of the lengths of the dry periods, just as for the flood sequence. The sequence of the appertaining mean annual rainfall can be also used for such an analysis.

One of the worst droughts in modern history was experienced during the period 1968 – 1974 in the Sahel region. Chad, Mali, Mauretania, Niger, Senegal, Upper Volta and Gambia suffered most. The drought extended to southern Sudan and Ethiopia. During the dry period, the Rivers Senegal and Niger fell to the lowest stages measured and Lake Chad evaporated to only one third of its normal size. The arid zone at some locations extended 100 km $year^{-1}$ southward. The soil infiltration also decreased due to the poor quality of the soil and so when heavy rains came in 1974, most of the water flowed away as surface runoff, causing additional damage to the population and crops.

7.4 REFERENCES

[1] Balek, J., 1972. An application of inadequate data for the Afr. trop. regions in engineering design. 2nd Int. Symp. on Hydrol., Fort Collins, Colorado, 8 p.

[2] Balek, J., Holeček, J., 1964. Contrib. to the calculation of flood probability. (In Czech). Vodní hospodářství, Prague, No. 4, 83—85.

[3] Batista, J. L., 1981. Calculo de gastos maximos de deseño utilizando observaciones hidrométricas. Voluntad hidraulica, XVIII, 24—26.

[4] Beran, M. A., 1979. Drought predictability problems and probabilities. Proc. Int. Symp. on Hyd. Asp. of Droughts. New Delhi, 229—241.

[5] Calp, M. M. Ed., 1942. Engineering Handbook. Soil Cons. Service, Milwaukee, Wis.

[6] Cervera, J. S., Arias, D., P., G., 1981. A perspective study of droughts in Mexico. Journal of Hydrology 51, 41—55.

[7] Chow Ven Te, 1962. Hydrologic determination of waterway areas for the design of drainage structures in small drainage basins. Univ. Illinois Bull., 462.

[8] Cordery, I., 1981. Risk and reliability considerations in hydrol. design. Hydrol. document for Hyd. Courses, Univ. of Roorkee, India.

[9] Dake, J. M. K., Design of spillways for small dams in Zambia. NCSR WR Rep. No. 12, Luska, 14 p.

[10] Dake, J. M. K., 1977. Development of arid and semiarid lands. Unesco, 42 p.

[11] Gibbs, W. J., Maher, J., V., 1967. Rainfall deciles as drought indicators. Australia Commonw. Bur. Met. Bull., 48, Melbourne.

[12] Greening, P. Ed., 1975. Proc. of a Symp. on Flood Hydrol., Nairobi. Transport Road Res. Lab. Rep. 259, Crowthorne.

[13] Gupta, S. K., Babu, R., Tejwani, K. G., 1970. Nomographs and important parameters for estimation of peak rate of runoff from small watersheds in India. Journal of Ag. Engineering 7, 25—32.

[14] Kovács, G., 1971. Relationship between characteristic flood discharges and the catchment area. Symp. on the Role of Hydrol. in Econ. Dev. of Africa, Vol. II, WMO Rep. No. 301, Geneva, 18—23.

[15] Krishnan, A., 1979. Definition of drought and factors relevant to specification of agricultural and hydrol droughts. Proc. Int. Symp. on Hydrol. Aspects of Droughts, New Delhi, 67—102.

[16] Krishnan, A., 1975. La recherche en zone sahéliene. Acad. des Sciences d'outre mer, NED, No 4216. Le Documentation française, Paris.

[17] Ledger, D. C., 1975. The water balance of an exceptionally wet catchment area in West Africa. Journal of Hydrology 24, 207—214.

[18] McMahon, T. A., 1979. Hydrological characteristics of arid zones. The Proc. of Symp. on the hydrology of areas of low precipitation, Canberra. IAHS Publ. No. 128, 105—123.

[19] McMahon, T. A., Srihanthan, R., 1981. Log Pearson III. distribution — is it applicable to flood frequency analysis of Australian streams? Journal of Hydrology 52, 139—149.

[20] Molina, C. S., 1977. Analisis hidrometeorologico de las inundaciones de los rios Quevedo, Vinces y Babahoyo. Inst. Nat. de Met. e Hidrologia, Quito, Ecuador, 27 p

[21] Monteagudo, D. P., 1978. Gastos maximos. Analisis comparativo del nuevo metod o de calculo. Voluntad hidraulica No. 45/46.

[22] Ogrosky, H. O., Mockus, V., 1957. The Hydrology Guide. Nat. Eng. Handbook, Sect. 4, Hyd. Supp. A. Soil Conserv. Serv., U.S.D.A.

[23] Pitman, W. V., Midgley, D. C., 1967. Flood studies in South Africa. Die Siviele Ingenieur in Sid-Afrika, 193—198.

[24] Rodier, J., 1961. Essais de determination des carasterictiques des averses décennales. Hydrol. Symp. Nairobi Proc., 121—126.

[25] Schiller, E. J., Ngana, J. O., 1981. Estimation of flood peaks with limited hydrol. data. Journal of Hydrology 51, 151—157.

[26] Schereschewsky, P. L., 1980. Forecasting drought and abnormal seasons. Water Res. Journal, Nov. 1980, 1—4.

[27] Slatyer, R. O., 1962. Climate of the Alice Springs area. In "Lands of the Alice Springs Area", N. Territory, CSIRO, Land Res. No. 6 Rep., 109—128.

[28] Starmans, G. A. N., 1972. Personal communication, Wat. Aff. Dept., Lusaka.

[29] Starmans, G. A. N., Shalash, S., 1972. High and Low Water Years in Zambia. Min. of Rural Dev., Luskaka, 67 p.

[30] Tabony, R. C., 1977. Drought classification. The Met. Magazine 106, 1—10.

[31] Wipplinger, O., 1961. Deterioration of catchment yields in arid regions, its causes and possible remedial measures. Hydrol. Symp. Nairobi Proc., 281—289.

[32] Yevjevich, V., 1970. An objective approach to definitions and investigations of continental hydrologic drought. Colorado Univ. Hyd. Pap. No. 23.

8 WATER RESOURCES

8.1 WATER RESOURCES PLANNING

The development of water resources is a significant part of the integrated development of many tropical regions. Therefore, there is a continual need for the comparison the demand for water and the water resources available. Most of the great tropical basins are allocated in two or more countries and thus water resources development is often an international problem. Different possible approaches which might be preferred by neighbouring states with different political systems make the adequate solution of water resources development impracticable in many cases. An initial stage of water resources development which should be performed jointly without much effort is the making of a water resources inventory of partial basins, large basins and hydroecological regions.

Work of this kind can be accomplished at various levels and in accordance with the various types of inventories already taken. It is true to say that such an inventory is never definitely accomplished. Among the many reasons for this at least one can be cited: new data from established and extended observational networks need to be processed, analysed and utilised in the most economical way. For instance, the monitoring of water quality has at the present time a very short history. However, the type of information supplied by the network should result in protective measures, the extent of which reflects the current situation and future trends.[1]

Three main problems can be encountered in planning various water development schemes:

a) The relation of water resources to the environment,

b) the search for the most rational way of using water resources,

c) the search for the most rational manner of their development.

For an adequate solution of the above problems there is a continuous need for the development, maintenance and operation of hydrological and hydrometeorological networks, groundwater observation, soil water regime observation and water quality monitoring. An assessment of available water resources and present and future demands are other problems which have to be considered in the initial stage.

When dealing with any of these problems in the tropics, the different conditions existing there have to be borne in mind, particularly the higher rate of energy flow through the tropical ecosystems, caused by higher temperature and greater potential productivity, the higher consumption of nutritive salts and thus a greater need for their replenishment. Waste products and dead organisms rot more rapidly and require a higher consumption of oxygen. A high evaporation and transpiration rate is another typical feature of the tropics.

An integrated approach should become a part of water resources planning because development should be planned in accordance with engineering, agricultural, technical, financial, economic, human and social aspects. The training of manpower, capable of further work in the development of water resources at all levels, needs to be a part of water development plans.

An adequate water administration at national, bilateral and international level has to be established for the efficient utilisation of water resources. Particularly important is the establishment of centralized institutions responsible for the collection and inventory of all data and for the coordination of various water projects and water rights.

In the planning of water resources development it should be remembered that the programme needs to be flexible with emphasis placed on priorities. Alternative solutions to the present problems under review should always be specified. According to past experiences it is strongly recommended to give priority to small and medium size development schemes and to projects in which results are most likely to follow immediately after their completion. Then a step by step extension of small pilot schemes can follow if the function of the pilot schemes has proved to be successful.

Self-help programmes with some supplementary assistance from governmental agencies should have priority and international assistance in terms of experts salaries, equipment and supplies should be considered as marginal.

Each plan has to be based on adequate data and if the data are not available a solid temporary network should be established as part of the preparatory work for each scheme.

Basic and applied research related to the specific problems of each scheme should be based on locally conducted experiments.

Problems of training, manpower and resettlement should also be clarified during the preparatory stage.

8.2 WATER SUPPLY FOR AGRICULTURE AND IRRIGATION

A very high proportion of the inhabitants of the tropics are engaged on the land as cultivators or pastoralists. In some areas more than 90 % of the inhabitants work in agriculture. Tab. 8.1 lists the ten countries in the tropics most dependent on agriculture. In most of these countries the variability of the climate is responsible for economic weakness. Without further improvement of water conditions total income from agricultural produce cannot be significantly increased.

In addition to irrigation, other factors which can improve existing conditions are shelter belts, the selection of more resistent crops and the proper timing of agricultural operations. However, any of these means is more or less dependent on water and water dominates tropical agriculture.

A very common feature of tropical cultivation is a shifting of the population known as 'ladang' in Indonesia, 'bewar' in India, milpa in Mexico and 'roca' in Brazil. Under such a system a plot of old forest underlined with fertile soil is cleared and then the brushwood is burned. Perfect timing of this work is essential because the brushwood has to be dried out before being burned but still no time should be left for fast-growing plants to regenerate. With the increasing uncertainty of the rainy season work and crops can easily be spoiled. Thus such a practice requires good natural knowledge of hydrological conditions and a certain ability to forecast them.

Tab. 8.1 Percentage of the population engaged in agricultural production

Country	Percentage of the population
Tanzania	92
Malawi	91
Laos	90
Bangladesh	86
Honduras	83
Thailand	81
Pakistan	75
India	72
Bolivia	72
The Philippines	66

The impact of burning has been discussed for a long time. Ecologists assume that frequent burning degrades the humus layer, however those with long experience in the tropics claim that the negative effect is not so significant. It has been estimated that as a result of the burning itself about $100-150$ kg ha^{-1} of nitrogen is lost in the atmosphere, especially when large plots are burned. On the other hand, phosphorus and potash from the ash enrich the soil.

The annual harvesting of quickly growing crops begins early during the rainy period, while later crops are harvested during the dry season. Clearance proceeds steadily moving outwards from the village until the distance reached demands another settlement. How does such a type of farming influence the hydrological regime? Some agriculturalists believe that because cultivation of this type is practised on soils of inferior quality, burning improves soil infiltration. Furthermore, control of weeds reduces evapotranspiration. However, so far, no real objective evaluation of the improvement is available.

Man adapts himself spontaneously to local conditions, nevertheless, such an adaptation can be very economical. Geddes [7] reported that the tribe of Dyaks on Sarawak grows rice upland under shifting cultivation instead of growing it in swamps, although in swamps the yield is 1900 kg ha^{-1} and in the upland it is less than 1600 kg ha^{-1}. However, as ascertained by Geddes, the yield in swamps is 0.87 kg per man-hour, while on the uplands it is 0,95 kg. Obviously the yield per unit of labour is more important to the tribe than the yield per unit area.

Wetland cultivation is widespread in south-east Asia and at present it is a type of farming very much encouraged in the humid tropics because the abundance of water and not the quality of the soil is the decisive factor. The negative aspects of such a type of cultivation are the requirements of grading and in hilly regions very laborious preparational work on the soil surface, such as terracing, is required. As well as an abundant amount of water, dry sunny periods are needed for the ripening of wetland crops, therefore the method has always been popular in monsoonal hilly regions.

Cash crop farming with a variety of crops for family consumption and also for the market provides entirely different conditions on many tropical soils. The selection of the cash crops reflects the particular environmental conditions.

Plantation is the final stage of the tropical land use pattern. While in the previous case local domestic plants are farmed, such as coconut palms in the Pacific area for copra production, on plantations induced-types of crops are grown and this in the long term changes the original hydrological cycle.

Classical examples are the plantation of rubber trees in Malaysia in the 19th century and the many attempts to change land use in swampy areas. The latter attempts have resulted in crop improvement on deep soils underlining the swamps, however, the acceleration of the water flow through formerly swampy areas has contributed to the destabilization of the flood regime in low-lying regions of basin. The destruction of the spongy effect of swamps releasing the water during the dry season also can have a negative influence on stabilized agricultural systems downstream.

Any assessment of the changing land use pattern should be accompanied, if not preceded by, in situ experiments conducted on several catchments. Some of them remain intact during the experimental period, other change suddenly to a new land use pattern. Such types of experiment are very laborious, expensive and not necessarily always successful, because the large scale effects may not be entirely identical with those achieved on a small scale. Pioneering work in this direction has been done in east Africa. Pereira and Hosegood described the results of experiments in Kenya in which indigenous bamboo forest was replaced by softwood types of Radiata pine and Monterey cypress. Because of the very similar root system and water being drawn from a depth of slightly over 3 metres, the evapotranspiration remained the same; the results indicate that a change of land use can be made without any danger of creating a water deficiency.

In the forested hills south of Lake Victoria much of the forest was replaced by tea plantations. For more than a decade water consumption was the same for both forms of land use, however, during dry years the consumption of water by the forest was less than by the tea plantation, although during the initial stages, when parts of the cleared soil were bare, consumption by the plantation was lower.

Also the restoration of overgrazed areas in Uganda was conducted with success and no damage to the hydrological regime was reported.

In contrast to shifting cultivation, 'paramba' type cultivation has been practised on the west coasts of India and Ceylon. The advantage of this type is that it does not upset the natural ecological balance because a variety of useful plants are grown simultaneously all the year round. Each paramba is a self-supporting unit and providing that the composition of the vegetation closely approaches the existing natural conditions, this kind of cultivation can be very successful.

Negative effects of the changes in land use resulting from the demand for cash crops such as rubber, pepper, tea, etc., have been reported from various parts of the tropics.

The present situation in the tropics indicates two extremes: One is a nostalgic approach on the part of environmentalists and conservationists while the second is a futuristic exploitable approach. From the hydrological point of view the approach should be realistic even if not always being half way between both extremes.

Eighty percent of water in the tropics is used for agriculture. Even so the population in many tropical countries is growing faster than the provision of food. An improvement in the water supply to the fields is one of the most effective means of increasing food production. Some reserves are foreseen in the better utilisation of monsoonal rains through more intensive infiltration of the rain, which otherwise flows away as surface runoff. The construction of water tanks collecting surface runoff or resettlement schemes can be regarded rather as emergency routines. Sometimes the situation can be temporarily improved by removing the phreatophytes and upland scrub and thus reducing evapotranspiration.

A considerable improvement can be achieved by effective irrigation. As can be seen from data of the F.A.O., the potential yield in many countries is much greater than the present yield. Data for India (Tab. 8.2) can serve as an example.

The effectivenes of irrigation can be judged from the following formula:

$$\text{Effectiveness} = \frac{\text{water placed in the root zone}}{\text{water located in the irrigation system}}$$

Potential losses which may reduce the effectiveness of any irrigation project are:

a) Seepage and evaporation from reservoirs. In the Sudan the total loss can reach almost eight metres (according to some sources), at Lake Nasser 2.5 metres and in India, 3 metres. As a means of protection the main methods used have been the application of oil films and shadowing vegetation, both with variable success.

b) Losses during water transport in canals. Protection can be achieved by plastic lining or by colmation.

c) Losses during distribution, which can be prevented by technical means such as hose pipe conducts etc.

d) Losses depending on the soil texture, soil moisture regime, root development and low infiltration. Protection can be given by the selection of proper irrigation routines such as sprinkling, by the calculation of the actual amount of water needed for irrigation etc.

Tab. 8.2 Average and potential yield of various crops in India

Crop	Average yield (tons ha^{-1})	Potential yield (tons ha^{-1})
Rice	1.6	10
Maize	1.1	11
Wheat	1.2	7.2
Tapioca	13.0	48.0

From the hydrological point of view the following effects resulting from irrigation should be considered:

a) increased evaporation and evapotranspiration,
b) possible changes of the rainfall regime in the microscale,
c) increased erosional and surface runoff potential,
d) groundwater regime,
e) migration of salts,
f) change of the regime in the unsaturated zone.

For irrigation itself three phases should be considered, namely source of water, transport of water and application of water. Often traditional practices can be found to be a suitable solution, provided they are well-adapted to the region. For instance, a tricky system of subsurface canals called ghanats in the Middle East and galerías in Mexico have proved to be economical (Fig. 8.1). Through these systems water is delivered from distant sources to the fields and transported by surface canals to the crops. Less developed routines are individual canals and/or furrows, diverting the water from the streams or main canals, the use of standing sheets of water as provided in paddy fields, flowing sheets of water as practised during and after floods, or the most simple transport of water using bucket. Almost always the ecological conditions have been the decisive factor in the selection of adequate traditional transport. Kirkby [13] provided examples of the environmental impact on primitive irrigation from the Mexican region of Oaxaca where different ecological conditions within a short distance of each other led to entirely different systems of water transportation.

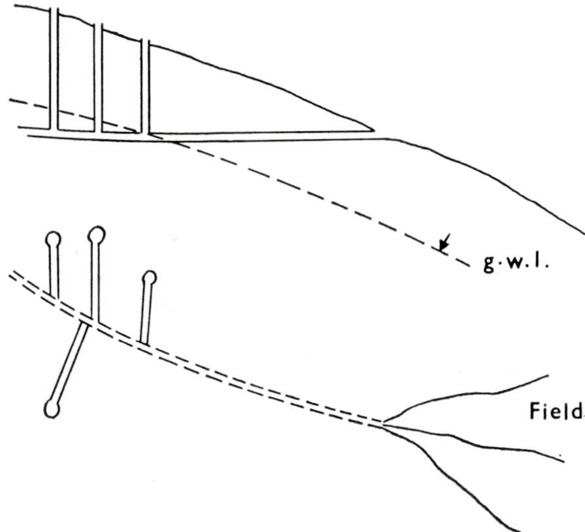

Fig. 8.1 Water transport through ghanats and galerías

Tab. 8.3 lists the tropical countries with the largest proportion of irrigated areas.

Tab. 8.3 Irrigated soils in the tropics

Country	Irrigated land (10^6 acres)
India	100
Pakistan	35
Indonesia	16
Mexico	11
Thailand	5.5
Australia	3.3
Peru	3.2
Chile	3.1
Sudan	3
The Philippines	2.2

Extensive schemes have not always proved themselves to be the best. Wortington [26] suggested that extensive schemes should not be formed and recommended a size between 20 – 300 hectares because if over 600 and under 40 hectares the soil often deteriorates. Other authors recommend that the optimum size should be 3000 to 5000 hectares with lower limit of 1000 and an upper limit of 10 000 hectares F.A.O. recommends medium-sized irrigation schemes on 40 – 4000 hectares. Big projects like the once-initiated diversion of the River Niger into the interior of the Sahara, have been proved risky and uneconomical.

The quality of water used for irrigation is equally important. In general, it is considered possible to use water with a salt content of $3-4\ \mathrm{g\,l^{-1}}$. Kovda [14] suggested the following relationship for judging the salt balance of irrigation systems, supplied from groundwater sources:

$$Yc = Q_1c_1 + Q_2c_2 + Q_{dr}c_{dr}$$

where Y is the amount of irrigated water at a specified time interval, c is the concentration of salts in the irrigation system, Q_1, Q_2 is flow of incoming/outgoing groundwater respectively, c_1, c_2 is the concentration of salts in incoming and outgoing groundwater, respectively, Q_{dr} is flow rate of the drainage water, c_{dr} is weighted groundwater salinity in a drained field.

Tab. 8.4 shows ten tropical countries with the largest area of soil affected by salts.

Tab. 8.4 Salt affected soils in the tropics

Country	1000 hectares	Country	1000 hectares
Australia	357 300	Chile	8600
India	27 100	Chad	8200
Paraguay	21 900	Nigeria	6500
Indonesia	13 200	Bolivia	5950
Ethiopia	11 000	Somalia	5600

As far as the amount of dissolved particles is concerned, silt particles 0.05 – 0.10 mm in size may improve the physical properties of heavy soils, while clayish particles improve soil absorption properties. However, they have a negative impact on permeability.

Quality is also judged by mineralization; less than $500\ \mathrm{mg\,l^{-1}}$ is considered to provide a good condition, $500-1500\ \mathrm{mg\,l^{-1}}$ a fair condition and more than $1500\ \mathrm{mg\,l^{-1}}$ a bad condition.

Good quality water for irrigation is when the ratio

$$\frac{\dfrac{Na^{+}\ \mathrm{mg\,l^{-1}}}{23}}{\dfrac{Ca^{2+}\ \mathrm{mg\,l^{-1}}}{20}+\dfrac{Mg^{2+}\ \mathrm{mg\,l^{-1}}}{12.16}}$$

is less than 1, fair quality 1 – 2 and poor quality 2 or more.

Also conductivity is used as an indicator of the quality of water for irrigation (Tab. 8.5).

Tab. 8.5 Conductivity as an indicator of the mineralisation and quality of water used for irrigation

Quality	Conductivity ($\mathrm{ohm^{-1}\ cm^{-1}\ 10^{-6}}$ at 25 °C)
Excellent	up to 250
Good	250—750
Acceptable	750—2000
Poor	2000—3000
Unacceptable	over 3000

Sometimes mineralization is used as a quality indicator in combination with the groundwater depth. For various depths of the groundwater available for agriculture, permissible values of the mineralilization are given in Tab. 8.6. However, the seasonal variability of both values in a single year should be taken into account.

Tab. 8.6 A critical depth of the groundwater for irrigation as related to mineralization

Permissible mineralisation ($\mathrm{g\,l^{-1}}$) ($\mathrm{g\,l^{-1}}$)	Groundwater depth (m)
0.1	1—15
1.5	1.5
2—3	2.5
3—10	3
20—30	3.5

The extent of irrigated agriculture is closely related to ecological zones and its intensification is recommended in:

a) arid lands where the settled agriculture depends on irrigation,
b) semi-arid lands to ensure safe yieldes,
c) areas with erratic rainfall for the production of high-value export crops,
d) areas suitable for crops with higher water requirements.

The first type of land is the most difficult to irrigate because the climate is too dry to permit successful growth of crops in an average year. Usually areas in the outer margins of semi-arid regions, semi-arid to arid areas and areas of extreme aridity are recognized. In the outer margins there is no risk of inadequate rainfall, in the second there is a chance of cropping in some years, in the third there is no rainfall throughout the year.

An estimate of water supplies for irrigation schemes should be made as a first step in the preparatory stage and the following areas are recognized:

a) Water is not available at all,
b) water is available in small local supplies such as pan valleys,
c) water is available only from intermittent streams,
d) water is available from areas with humid climate,
e) water is available from groundwater sources,
f) water is available from only ocassionally replenished streams.

In many regions drainage is the most important problem. Flooding, waterlogging and increased salinity are among the problems which can often be solved by drainage and flood-control techniques.

Fisheries are an equally important source of food from tropical rivers and lakes. Both are very rich in the variety of species. Roberts [20] found 260 species of fish in the River Kapuas, Borneo, some 30 of which have been scientifically described only recently. In this respect, tropical fresh streams are still among the world's great unexplored areas.

Although fish is not looked upon as a staple diet by many tribes, practically all tropical rivers and lakes are fished at some time or other. In monsoonal regions it has been found possible to produce 250 – 500 kilograms of carp per hectar. Even in small ponds or intermittent streams carrying water for about nine months of the year, annual commercial production could be 8 – 10 times higher than in European commercially exploited ponds.

White estimated that in the irrigation systems the production of fish may reach 400 kilograms per hectare. On paddy fields in Malagasy 20 – 30 kg ha^{-1} has been produced in 120 days and 80 – 200 kg ha^{-1} in the same time, after the application of fertilizers.

Rivers are a significant source of fish even in countries with an extensive coast. In Indonesia 36 % of fish comes from rivers, in India 28 %, in Pakistan 56 % and in Kampucheja 70 %. In artificial lakes production is much smaller than in natural ones. In Kariba, for instance, the total catch decreased by six times in two years.

The situation requires the establishment of artificial pilot schemes and the continuous improvement and introduction of new species.

The amount of fertilizers suitable also varies from place to place. The establishment of small shallow ponds has been proved more effective than the utilisation of deep reservoirs.

The over-exploitation of water resources, particularly in the highlands, has caused a decline in various species of the fish population. The spawning runs of some species in Ethiopian rivers have been reduced to fractions of their former abundance. It has to be borne in mind in any water management scheme that the transformation of a stream from fast flowing rapids to deep still water, or vice versa changes one ecosystem into a quite different one and so eliminates many migratory and sedentary species.

The role of the hydrologist is essential during the preliminary stage of project planning, in seeking for the best locations for ponds and determining the duration and seasonal distribution of the flow of water.

8.3 RURAL WATER SUPPLY

One of the main problems of tropical villages, particularly in semi-arid regions, is the provision of daily supply of water. The author observed women in Gwembe Valley near the Kariba reservoir, carrying daily supplies of water for a distance of 12 kilometres. A similar observation was reported from Yucatan villages in Mexico. Omari [17] found in the region of Kilimanjaro that the village women spent up to ten hours a day bringing water to the villages.

It has been calculated that an active man's daily requirements of drinking water alone often exceeds 10 litres in hot deserts and may rise to 20 litres or more. The minimum needs of water for domestic and hygienic purpose must be added to that amount. Bearing in mind that often much of the water readily available may have too high a concentration of mineral salts for use without treatment, the seriousness of the problem can be appreciated. Fig. 8.2 shows daily water intake in relation to air temperature and the level of the day's activity.

Nevertheless the criteria for local supply potential vary greatly. A bushman in the Kalahari needs only one litre per day while an inhabitant of an industrial tropical city may consume up to 800 litres per day.

In general, two ways of increasing the water supply are recognized in arid and semi-arid regions:

a) Through improvement of the existing hydrometeorologic conditions,

b) by improving the possibilities of making full use of the existing water supply potential.

The list of tropical water resources which can be utilised for the rural water supply includes:

a) surface streams which maintain water all the year;

b) subsurface streams found at locations with gentle slopes, where the water moves slowly into the stream;

c) phreatic water which is normally thought to be of poor quality being saline, but which is easily obtainable from shallow wells and boreholes in coastal dunes;

d) artesian water that can be pumped or which flows up to the surface;

e) springs, which frequently supply only limited amount;

f) artificial tanks collecting torrential water.

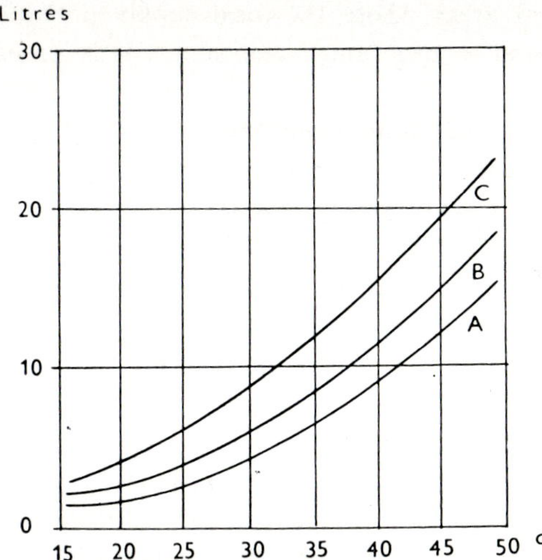

Fig. 8.2 Daily water intake recommended by U.S. Army in relation to air temperature and level of day's activity. A, rest in shade; B, moderate work in sun; C, hard work in sun

In general, it must be taken into consideration that some sources supply water at the expense of other regions and frequently it is water which cannot be replenished during our era.

Seven types of regions can be found in the tropics which differ in water supply potential:

1. Extremely arid regions can be supplied only through long-distance transport of water.

2. Sparse localized supplies are found in areas with 100 – 400 mm of annual rainfall. Some water can be collected in localized areas. Surface flow is ephemeral, erratic and violent in form. The water seeps from stream channels into valley sediments and evaporates in shallow playas. Small basins of groundwater supply and river beds are replenished and water is locally available.

3. Semi-arid supplies, found in regions where the seasonal flow in the rivers varies from year to year, but where some perennial streams are available as well as

underground sources. Sufficient precipitation supports the crops in some years and there is a possibility of improving the supply by the construction of small dams and by tapping shallow wells.

4. Exogenous surface water supplies are found at locations where extremely arid areas are crossed by major rivers with a perennial flow forming a high-potential groundwater supply.

5. Exogenous groundwater supply regions are in arid or semi-arid areas where the underlying aquifers are replenished from humid or semi-humid areas.

6. Localized exogenous supplies are found in areas where a sharp difference in relief and rainfall exists over a short distance. Arid valleys are fed by small aquifers from mountain chains in their vicinity.

7. Endogenous supplies are found in areas where the local supply potential exceeds the actual evapotranspiration and seepage and water supply is in excess of the need of crops and human beings.

Rural water schemes are normally not self-supporting and have to be built with governmental assistance. Even non-expensive schemes can improve living and health conditions significantly. An extensive programme for the conservation of water should accompany any long-term planning of water resources utilisation. Careful application of farming practices, soil fixation through rapidly growing plants, terracing, contour cultivation, grazing control, afforestation, etc., are simple, inexpensive means which should be developed and supported by the administrative organs.

Systematic appraisal of water resources, reconnaissance survey, intensive observations in experimental and representative watersheds selected in ecologically different regions, systematic silt sampling and measurement of the sediment load of the rivers are the main tasks of hydrologists in the field of water resources development.

8.4 THE ENVIRONMENT AND POLLUTION

Environmental quality and its current deterioration is entirely man's responsibility. The contamination of the natural environment can be accidental, resulting for instance from large fires, the release of industrial water and effluent, the abuse of explosive substances, exhaust from vehicles, from sewage; deliberate as from the testing of nuclear devices and agricultural and industrial burning; and purposeful, resulting for instance from the application of fertilizers, pesticides, hebicides, defoliants, etc.

Closely related to contamination problems is the growth of the population and relevant problems of housing, feeding, health, etc. At the end of this century the world's population will be approximately twice what it is today and 80% of the total population will be living in tropical developing countries. Urbanization, industrilisation and increasingly significant chemical contamination arising from industrial and agricultural development will be features accompanying this increase. In less developed countries herdsmen and subsistence farmers will move out to the marginal areas of semi-arid regions and through the over-use of the land environmental deterioration may occur. This is already being witnessed in some parts of Africa, Central America and the Caribbean islands.

Eutrophication of natural waters, their microchemical pollution by synthetic organic substances, detergents, phenols, thermal pollutants, hydrocarbons, nitrates etc., are other phenomena of the deterioration of the environment. Although artificial pollution of natural resources in the tropics is still rare and limited to the vicinity of big towns and industrial centres, occasionally we can find tropical streams polluted near villages by the foam of detergents.

A great hazard exists in the biological pollutants found traditionally in tropical streams and in some cases artificially induced as in the vicinity of irrigation canals.

Open irrigation can transmit cholera, typhoid, dysentery, hepatitis, pliomyelitis and bilharzia. Canals are also ideal media for hookworm, eelworm, larvae and amoeba cysts, schistosomiasis and malaria.

Standing and slowly-flowing waters are favoured by mosquitoes, transmitting malaria, filariasis, yellow fever and encephalitis, by simulium flies transmitting onochcerciasis, by tse-tse flies transmitting trypanosomiasis, by snails transmitting bilharzia, paragonimiasis and distomatosis. Therefore health specialists should always be included in irrigation project teams in the tropics for the preparation of protective measures.

Bathing and swimming in tropical waters can be very dangerous. Warm abundant supplies of water in which mosquitos and snails breed, are sources of infection. It is not only shallow muddy water which is favoured by mosquitoes. In the Red River delta, in southern Asia an absence of mosquitoes was observed because they cannot tolerate rice fields. Upstream in the hilly regions with clear running water the region became highly malarical after the felling of trees which had shaded formerly cool water.

Soils can be easily damaged by careless pollution. In addition to salinized soils listed in Tab. 8.4, fertilizers, pesticides, industrial waste, concrete, bitumen, metal residues and radionuclides have negative and cumulative effects on soils which were once fertile. Relevant to this is food contamination particularly through the slow biodegradibility rate of DDT and DDE.

In many tropical countries the main problems of environmental protection are still seen in nature conservation and game protection. (Putney et al. [19]). Environmentalists frequently oppose any possible threats of dams, irrigation schemes or agricultural development. However, on the other hand environmental damage may be caused by unreasonable conservation policies. At one time, the Luangwa Game Reserve on the Zambia – Malawi borders became so overherded by elephants that wide areas were seriously damaged and lost their vegetational cover exposing the soil to intensive erosion. This, of course, had a negative influence on the hydrological cycle.

In slowly moving shallow waters vegetational pollution is another serious problem. The water hyacinth and other species are widespread in these waters and many protective methods including spraying from the air have been tested with limited success.

Solid wastes may also become a hazard to surface and ground water. Rossi [21] developed an effective method for the tropics of converting putrescible waste through the planned systematic use of earthworms. The method has been successfully applied in the Philippines.

The quality of tropical surface water must always be regarded with suspicion, particularly at the end of a dry season, when concentrations are much higher. If regular water treatment is not possible, sand filters, tubewells in the banks or infiltration galleries should at least be constructed. Many sources of surface water though harmless to the local population, may provide a source of disease for people coming from the outside of the region.

Industrial activity is becoming a more frequent source of water pollution in the tropics. In addition to pollution by metals, the salinity of water resources is increasing, especially in mining areas. Mann and Chaterji [15] reported such a type of pollution from Rajasthan, India and recommended the use of gypsum as a soil amendment and the introduction of various plant species for the control of salinization.

A method of water treatment in the tropics was developed by Frankel [6]. It is cheap and locally available materials can be used. Coconut fiber and burnt rice husks can form a two-stage filtering gravity process. The author claims that in most cases there is no need for the addition of chemicals. The coconut fibre acts as a substitute for coagulation and sedimentation while the rice husks behave like any sand filter, removing taste, colour and odour through the absorption properties of activated carbon. The system has been successfully applied in the Philippines and Thailand and proved to kill 60 – 90 % of bacteria. Thus only a trace of chlorine was required to achieve water quality in accordance with W.H.O. standards.

The results of the survey on optimal waste water treatment by standard methods with recommendations on the most suitable methods for developing countries were reported by Ellis [5].

Groundwater sources, except for shallow wells, which in any case are traditionally avoided by many tribes, are much safer. Again chlorination is a simple method even if it is not always a substitute for filtration.

Tab. 8.7 shows desirable standards for drinking water and for water for domestic use, as recommended in Gambia. They can serve as guides only because the desirable standards may vary from country to country and from time to time, reflecting the stage of local development.

Similar standards have been set up for various types of fertilizers, pesticides and herbicides. The standards are still very variable and it is beyond the scope of this book to describe them. In general, it has been emphasized by various authors that

Tab. 8.7 Recommended water quality standards for drinking and domestic use

Substance	Permissible limit (ppm.)	Substance	Permissible limit (ppm.)
Alkalinity	200	Iron	0.3
Arsenic	0.01	Lead	0.05
Barium	0.5	Magnesium	50
Boron	0.5	Manganese	0.2
Calcium	50	Mercury	0.005
Chlorides	200	Nitrate	20
Chromium	0.01	pH	7.0—8.5
Cobalt	0.3	Phenol	0.001
Colour Hazen scale	10	Sodium	50
Copper	1.0	Sulphate	200
Cyanide	0.01	Suspended solids	40
Dissolved solids	500	Uranium	0.5
Fluoride	1.1	Zinc	1.0
Hardness-total	250		

the simple use of artificial fertilizers in the humid tropics can give very disappointing results unless the soil environment is particularly favourable. Soil structure, together with the water in the soil are of the greatest importance. One crop benefits from one type of fertilizer and another from a different one. The main types used in the tropics are those which supply nitrogen, phosphates and lime, while sulphur, magnesium and iron are frequently widespread in tropical soils.

The effect of nitrogen application on cereal yields in India, based on F.A.O. records, is given in Fig. 8.3. An increase in application does not result in an adequate yield, however, the negative impact on groundwater pollution rapidly increases.

The negative effect of tropical regions on conditions of life is very pronounced. Constant heat and humidity sap most of man's vitality. Huntington [10] stated as early as 1915 that the optimum temperature for physical work was 18 °C when accompanied by 75 – 80 % humidity. Taylor regarded a temperature measured by a wet bulb thermometer of between 7 and 15.6 °C as the optimum and believed that it rarely resulted in uncomfortable conditions. Above 18 °C conditions were considered as frequently uncomfortable and above 21 °C as continuously uncomfortable.

Climatic effect is an equally significant factor. The then U.S. Weather Bureau [23] introduced a Discomfort or Temperature – Humidity Index (THI):

$$THI = 0.8T + \frac{RHT}{500}$$

where RH is relative humidity (%) and T is the air temperature (°C). This indicates the temperature, which combined with relative humidity of 100 %, would create the same reaction as the existent combination of temperature and humidity. At THI = 21 most people feel comfortable, at 24 they feel thermal stress and at 26 almost everyone experiences discomfort.

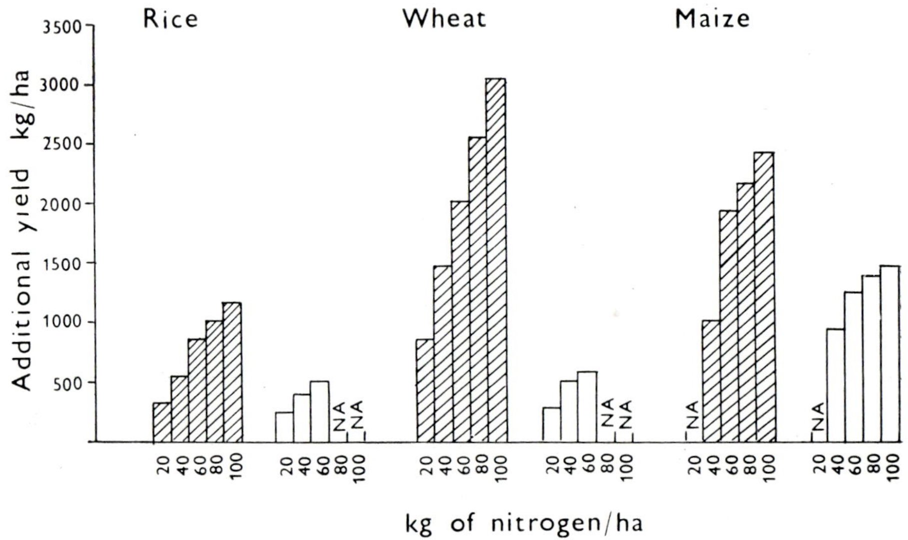

Fig. 8.3 Effect of nitrogenous fertilizer upon cereal yields in India. Hatched columns indicate hybrid, white columns indicate local variations. N.A. — data not available.

It is not only temperature and humidity which cause a lower degree of physical and mental ability. A pronounced dry season, particularly the final part of it and the need for quick adaptation to entirely new environmental conditions, which may exist from region to region, arouse various degrees of nervous strain.

As stated by Verboom [25] the tropical environment is also increasingly damaged by visible pollution. The countryside is intensively littered with empty plastic bags, containers, glass, waxed cardboard, wrecks of old cars, etc. Apart from costly measures, wide-scale health education and the promotion of antipollution sentiments are effective ways of combating pollution and protecting the environment. Various possibilities of such an approach were widely discussed by Pickford [18].

It can be concluded that tropical resources may be reasonably developed without drastic deterioration, providing that basic protective measures are included in the initial stages of planning. An example of such an approach on a continental scale can be found in special publications of the United Nations Development Program, concerned with water management and the environment in Latin America, Asia and Africa.

8.5 LARGE SCHEMES, ENERGY AND NAVIGABILITY

Thanks to relief and climate there are great possibilities of hydroelectric power production in many parts of the tropics. However, abundant rainfall throughout the year decreases in a poleward direction with increasing distance from the equator. This has to be considered when reservoirs are being designed. As far as relief is concerned, the ancient Precambrian massifs of Africa, India and Brazil provide numerous sites for dams on their edges, where rivers cascade towards the sea or inland where the rivers flow through steep gorges or fall from one plateau to another. In one of its stretches the River Congo descends almost 300 metres through 32 falls and cataracts over a distance of 350 kilometres. The river develops a potential of 83 million kilowatt, at least 15% of world power potential.

Tab. 8.8 Water energy on the continents

Continent	Available 10^3 MW	% of world capacity	Developed %
Africa	780	27	2
South America	577	20	5
Southeast Asia	455	16	2
Australia	45	2	2

Tab. 8.8, compiled from various sources, indicates the power capacity which can be produced by river water energy. Tab. 8.9 shows the largest power schemes of the tropical world. A comparison of the two tables indicates that the power capacity which can be produced by water energy is far from being fully utilised.

The future situation for water power development is primarily a question of cost per unit, of the availability of capital and of the possibility of transmitting the energy over long distances. A significant advantage of water schemes is the absence of pollution and relatively easy maintenance.

The majority of large water resource schemes and dams in the tropics are multi-purpose. Various attempts to develop super multi-purpose schemes in the tropics on a continental scale have been witnessed on various occasions. At the beginning

Tab. 8.9 Largest energy schemes

Scheme	Country	Output MW
Itaipu	Brazil, Paraguay	12 600
Solteira	Brazil	3200
Guri	Venezuela	1400
Kariba	Zambia, Zimbabwe	1400
Jupia	Brazil	1400
Cabora Bassa	Mozambique	1200

of this century a project for the diversion of waters of the Niger into the central Sahara region was proposed. Later a Niger Office was established under which the Diamarabougou Dam was built with the aim of re-irrigating land once flooded by the old channels of the Niger, which used to flow into the vast lake of Timbuktu. Apart from technical difficulties the population was not prepared for such a large-scale resettlement.

Plans to divert the Congo into Lake Chad by flooding about 10% of Africa by what was to be called the Congo Sea were made with the aim of irrigating parts of the Sahara. A major canal for this purpose was planned between Lake Chad and the Mediterranean. The Ubangui ridge and the Sahara relief were found to be the major physical difficulties in addition to political and sociological ones.

In the Kalahari, the restoration of the inland lake was recommended by the weiring of the Cunene and Chobe rivers, providing an expected rainfall increase of 250 mm.

The general scheme of the Upper Nile Project was based on the formation of century-storage reservoirs and an attempt to protect water from evapotranspiration in the vast swamps of southern Sudan. Today the River Nile with seven dams and additional barrages is among the most exploited rivers of the tropics. Aswan Dam itself has an installed capacity of 2.1 million kilowatts which covers 53% of power requirements in Egypt.

Many problems are related to the construction of high dams in the tropics, one of the most serious is siltation and erosion. As for the Aswan High Dam, the amount of annual silt storage in the reservoir is 60 to 180 million tons (Manly and Hafet [16]). Thus it has been estimated that the dead storage of the reservoir will be filled within 500 years and the whole reservoir in 1400 years.

The reduced amount of silt deposited in the Nile delta, led to the steady sea erosion of the delta coastline along both the Rosetta and the Damietta peninsula.

The chemical and biological balance of rivers is also influenced by the construction of large reservoirs and a considerable amount of water is lost due to evaporation from the increased surface of the water. Transformation of riverine ecological conditions to lacustrine conditions, combined with the seasonal effect of the storage of floodwater result in various types of thermal and chemical stratification, the effects of which are not always easily predictable.

Fish distribution is almost always influenced by changed river regime. Public health conditions can be improved or worsened. Problems of resettlement arise when large schemes are under preparation, as happened during the construction of the Kariba Dam on the Zambezi, where people had been living in the fertile valley for hundreds of years. In many cases the rural population leaves for the towns and becomes less productive and more dependent on governmental welfare.

Therefore large schemes are more often considered as remote from reality and a preference is given to small and medium-sized schemes. Smaller schemes become effective in a much shorter time and can be more easily managed and controlled with local manpower.

One of the positive factors of large reservoirs is that they facilitate navigability. Water transport is almost always low-cost and in many parts of the tropics it plays significant role in the transportation system. This is particularly valid in the lowlands or on the plateaux where the rivers flow smoothly. Typical examples can be found in northern Brazil, Nigeria, Zaire and Burma. In many parts of wet and dry regions rivers are not navigable throughout the year. A typical example is the Niger—Benue river transport system. The upper Benue is navigable for heavy vessels for only two months of the year.

Most of the navigable rivers are short and with ungauged courses. Owing to heavy surf and shifting sands connections between oceans and the interior are very limited. Of the tropical streams the most suitable for heavy vessels is the Amazon which with tributaries is navigable 3600 km from the coast. Almost half the route has a depth of over 30 metres. Nevertheless the navigable potential of the basin has not been fully developed.

In some cases the remoteness of the country makes it difficult to navigate the river. An example of this is the Orinoco in Venezuela. Some rivers are navigable only in stretches, because of the numerous rapids. The Congo system has 13 000 kilometres of navigable waters and almost 3500 of them can be traversed by large vessels. However, the navigable section of the main river is divided into five parts.

In the search for sources of power in the tropics solar energy has been examined.

It has been calculated that in tropical areas with bright sunshine a power station with an output of 50 MW can be constructed on an area of 2.5 km^2.

In the tropics the difference between the temperature of surface water and water 1000 metres below the surface may amount to 4.4 °C. The U.S. Department of Energy financed a project called OTEC – Ocean Thermal Energy Conversion – which examines the possibility of constructing large power plants that would utilise this temperature difference. The generated electricity can be transmitted ashore by cables or used to extract hydrogen from seawater as a fuel. Off the coast of Hawaii experiments have been carried out with the extraction of solar energy from the upper layers of the ocean. It is anticipated that large parts of the tropics can become a source of energy for other parts of the world also.

8.6 WATER LAW AND INTERNATIONAL RIVER BOUNDARIES

Existing water legislation in many tropical countries derives from one or a combination of more than one of the following legal systems:

a) customary law;

b) Moslem law;

c) Roman law with two main derivations, namely Anglo-Saxon and "Code Countries" laws of French, Dutch, Belgian, Spanish and Portuguese origin.

Customary laws vary from country to country. They cover land ownership, cultivation, watering and fishing rights. Private ownership of water was generally unknown and individuals had the right to use water, while land tenure was communal or tribal.

According to Moslem law, water entails a religious obligation derived from its nature and no one can refuse to give surplus water without sinning against Allah and man. All members of a Moslem community are ensured of the availability of water and all waters are deemed to be common property. Since the principles of Islamic law are still strictly followed in tropical Moslem countries, any modernization of water administration must take them into account.

Roman law divided water into three categories, private, common and public. Private water, springing or flowing within one's property, is part of that property, common is any water not owned by anyone, such as rain, the seashore and non-navigable rivers. Public water includes large rivers, canals, lakes and is governed by public law.

In the absence of a centralized water organization there are usually many ministries or departments directly or indirectly responsible or interested in purely sectorial aspects of water resources.

At several meetings organized by the United Nations, problems which in many countries make water legislation inefficient were specified:

a) complexity;

b) non-acceptance of modern management practices and techniques;
c) undesirable fragmentation of administrative responsibilities;
d) dispersion of legal norms in different legal bodies,
e) contradiction between legal norms of national level and those of local level;
f) incompatibility between customary and traditional water rights and the role of the state.

Cano [3] pointed out other problems such as insecurity, rigidity, uncompleteness, lack of functional coordination, lack of territorial integration, orientation of the law on use instead on protection and lack of realism.

Obviously there are no uniform models of law that can be applied in all countries and thus the interchange of experiences is valuable so that they can be adapted. At the U.N. Water Conference at La Plata in 1977 it was recommended that water legislation should include:

a) public ownership of water and of the large engineering water works,
b) land ownership problems,
c) litigation,
d) protection of the quality of water,
e) prevention of pollution,
f) penalties for undesirable effluent discharge,
g) granting of water rights,
h) abstraction,
i) terms of reference of governmental agencies.

As stated by Burchi [2], in semi-arid and arid common law countries, such as Australia and South Africa, water rights have been made increasingly subject to governmental control. Under such a law public ownership of waters is vested in the state and public waters form part of the state's public domain, from which they are leased back for diversion, abstraction and use by individuals and communities. Also some non-arid countries such as Panama and Colombia have already placed all water resources both surface and underground under the state's public domain. In other countries there is a trend to centralise all aspects of water management, as has already been accomplished by Mexico, Venezuela, Kenya and Mauritius. In the Philippines the coordination of interdepartmental responsibilities is made through the National Water Resources Council.

In many countries, however, problems stem from the conflicting interests of several governmental organizations and ministries. For instance, water for irrigation is under the control of ministries of agriculture and irrigation, power plants belong to the ministries of energy, pollution is dealt with by health ministries, etc.

Thus from the point of view of integrated water resources planning and development and relevant aspects of water law, the following aspects of water legislation should be tackled at various stages of planning:

a) contribution of all activities to rational conservation, development and the utilisation of water resources,

b) inter-relationship with other natural resources and other environmental elements,

c) definition of national elements with respect to water,

d) gathering of information through individuals and public bodies,

e) control and ownership of water,

f) reservation of certain sources,

g) priorities between regions,

h) conditions of use,

i) legal system affecting other persons and their property with respect to water use,

j) judicial aspects of water,

k) participation of users and other persons,

l) water resources administration,

m) cadaster of waters, water use and water rights,

n) evaluation and approval of new technologies.

The international aspects of water law are equally important. Problems of international river boundaries occasionally become a matter of hydrological interest. Legal problems are mainly concerned with definition and demarcation. The boundaries usually follow:

a) The shore,

b) the median line,

c) the thalweg,

d) an arbitrarily selected line.

The first type of boundary is rare and an example is from Shatt el Arab where the boundaries between Iraq and Iran drawn in 1914 along the low water line limiting Iran's free access to the river became a source of conflict. The second type of demarcation line could entail great inconveniences because it changes its position with the fluctuation of the water level as happens on the Zambezi border between Zambia and Zimbabwe.

The thalweg or line connecting the deepest parts of a river channel is commonly used. However, in some treaties the thalweg and the median line were assumed to be coincident. This may be true in the straight part of the river channel, far less where it meanders. For instance, the border between Zambia and the then South-West Africa was defined as a thalweg and the difference between the thalweg border and the previously mentioned median line border at the theoretical point where the borders of Zambia, South-West Africa, Botswana and southern Rhodesia met, actually disconnected the boundaries between South-West Africa and Southern Rhodesia (Fig. 8.4).

An example of the fourth type is the latitude parallel on Lake Victoria between Tanzania and Uganda.

The shifting of river channels after flood or by silting may become a matter of serious dispute even on a large scale. An example can be taken from the area of almost 15 000 km^2 between Guyana and Surinam in the triangle between the Courantyne River and the so-called New River. While Guyana claims that the New River is in fact a tributary of the Courantyne, Surinam claims that the New River is in fact a continuation of the Courantyne set up as a boundary, and wrongly named New River on the original maps.

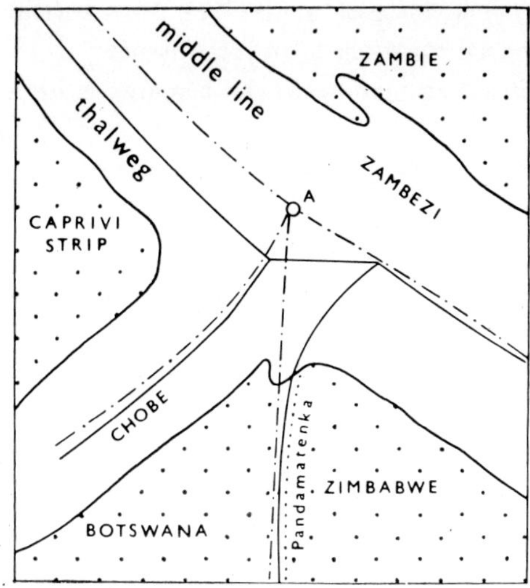

Fig. 8.4 Disconnected boundary at the point where the borders of Zambia, Botswana and the then South-West Africa and southern Rhodesia met

In such a controversy if not solved by military means, a decision as to which is the main stream has to be made by a hydrologist and needs to be based on the judgement as to which stream is longer, which has greater discharge, which has a greater drainage basin and greater geomorphic age and on the origin of the river beds above and below the junction.

Earlier treaties on tropical international rivers were concerned with navigation and boundaries. Recent treaties put more emphasis on economic development. Since lack of data is one of the first difficulties encountered in these studies, cooperation among states should start with the collection and exchange of data. Thus ideal international cooperation literally starts with meteorology and hydrology.

8.7 DATA AND DATA DEFICIENCIES

With the exception of the Nile, hydrological data in the tropics are very scarce. Only during the International Hydrological Decade were a denser network of observation posts established and data collected and analyzed. However, the records are still relatively short. Both meteorological and hydrological data are essential for the effective management of water resources and for the proper operation of water resources schemes already accomplished. For the temperate regions many methods have been developed, particularly in the field of stochastic hydrology, based on the utilisation of both historical and recent records. These methods cannot be used effectively in the tropics because of the lack of historical records. Instead of this modified methods based on short term records need to be developed. The continuous extension of observational networks remains a high priority.

Bearing in mind the length of records, Tschannerl [22] recognized three types of data in developing countries:

a) Historical data which should be utilised whenever possible. In practice this means the full utilisation of long rainfall records, most frequently available, for the extension of streamflow records.

b) Data which have not previously been widely utilised and which can be processed without delay for the special purposes of water resources schemes. Data obtained from tree rings, mud, varves and data obtained through methods of isotope hydrology are in this category.

c) Data collected at the cost of postponing the initiation of water resources schemes, or in other words data from new networks established during the preparatory stage of the projects.

In vast tropical areas the economical aspects of the observational network have to be considered and therefore the number of observation stations that require indefinite periods of observation should be kept within reasonable limits. Therefore preference is given to stations operated on a temporary basis. The intensive maintenance of these stations requires several years. Then the data can be analysed and extended by a comparison with the findings of the principal observation stations. Special phenomena observed for various specified purposes important for the given project are collected from special stations which carry out observations for very short period of time. Thus it is possible to differentiate between:

a) principal stations carrying out observation for an indefinite period;

b) secondary stations carrying out observations for a period of several years;

c) special stations concerned with special phenomena and making observations for a very short period.

The problem of network density has become a matter of interest in relation to climatic and orographic factors. The density is largely determined by climate, topography, geology, vegetation, population density and economic development. The purpose for which the network should be established is another important factor. For a general evaluation of water resources a less dense network may be sufficient, while for special water problems and research a temporary but very dense network can be found more suitable. The World Meteorological Organisation established the criteria for the network density in accordance with the topography and climate. The criteria vary in accordance with observed phenomena.

Precipitation stations should be most dense, because rainfall records are essential for almost any type of appraisal of water resources. The variability of precipitation should be well evidenced by the established network and therefore at least 10 % of the rain gauges should have recording facilities.

Next to precipitation, stream gauging stations supply the most valuable hydrologic data. The network should be planned so as to ensure an adequate sampling of the hydrologic variability within the region and characterize floods and droughts. It is recommended that gauging stations should be established below the junctions of main tributaries, above river mouths, at national borders, above diversions of irrigation canals and at outlets from lakes.

Groundwater observations should be established primarily in areas of significant groundwater development potential and should represent the probable extent, thickness and characteristics of the aquifer. The data should provide information on the seasonal variation of the groundwater storage, groundwater replenishment and safe yield. In addition to groundwater level observations, the drilling of test wells, the provision of pumping tests and geochemical observations are required.

Water quality measurements are significant at various stages of projects. The monitoring of water quality should be established at strategic locations and at existing gauging stations. In general, the water quality network in arid regions should be denser than in humid regions, because the concentrations are much higher in arid conditions.

Observation of sediments is often required above reservoirs for navigation purposes and in the vicinity of harbours. Two phases are recognized in sediment observation, bed load and suspended load. Typically they occur simultaneously and both phases should be observed. Suspended load data consist of periodic samples of the stream flow containing the suspended load at different water levels and bed load samples are collected at selected time intervals.

The measuring points should be selected only at stream gauging stations, so that the rating curve showing the relation between discharge and sediment load can be derived.

Evaporation data are needed in water balance studies and for the assessment of water loss from reservoirs. The measurements are based on pan measurements. The need for evaporation data increases with the degree of aridity. Usually, pan evaporation is established at each meteorological station. The establishment of representative observations suitable for regional comparison is difficult in view of the many types of pans, and they provide slightly different results.

Tab. 8.10 shows standards of network density as recommended by the W.M.O. Two types of network are considered, a minimum network and a provisional one, which indicates that further network extension is foreseen.

Tab. 8.10 Network density as recommended for tropical regions

Region	Minimum network	Provisional network
	area in km² per station	
Precipitation		
Flat	600—900	900—3000
Mountainous	100—250	250—1000
Mountainous islands	25	
Arid	1500—10 000	
Stream gauging		
Flat	1000—2500	3000—10 000
Mountainous	300—1000	1000—5000
Mountainous islands	140—300	
Arid	5000—20 000	
Groundwater	15—25	
Water quality		
Humid	5% of stream gauging stations	
Arid	25% of stream gauging stations	
Sediment		
Humid	15% of stream gauging stations	
Arid	30% of stream gauging stations	
Evaporation		
Humid	50 000	
Arid	30 000	

Not only the density of the network but also the length of records is significant in water resources planning, especially for the evaluation of the probability of the occurrence of extremes. It has to be born in mind that the old records are of a poor quality and not always fully compatible with the new ones, with regard to the development of new methods and instruments. So far a minimum period of observation has been recommended only for precipitation, however, the recommendations greatly differ in the length of the period.

It should be noted here that a difference between arid and humid tropics does not necessarily mean a difference between rainy and non-rainy tropics. Gourou [8] showed that both Dakar, where the rainy season lasts for over three months and annual rainfall is 521 mm and also Aden with 63 mm of annual rainfall, are regarded as humid tropics. The presence of humidity in the air does not necessarily mean the occurrence of rain. Seasonality is another important factor in addition to the total amount of rainfall. A previously cited author insisted on the importance of seasonality giving Bombay as an example with 2032 mm of annual rainfall falling within five months and Manaos with 1776 mm of annual rainfall and no pronounced dry season. These facts should be also considered before any rainfall network is set up and the same is valid for the observation of other phenomena.

Not only data collection but also data storage is significant. Checked and corrected data should be accessible in a form most convenient for future use. Provided they are to be used for further processing on a specified type of computer, the computer staff should be consulted before any type of databank is developed. Theoretically, data stored on tape or cards can be processed on any type of computer, however, the initial problems of data utilization should always be foreseen. In some cases standard data processing in the form of yearbooks and varried out more or less manually can be found more convenient, particularly when the data are less numerous and collected during a short period.

Special attention has been given in tropical regions to the establishment of experimental and representative basins. An experimental basin is one which has been chosen and instrumented for the study of hydrologic phenomena under changing conditions while a representative basin is one that has been chosen and instrumented to represent a broad area in lieu of making measurements in all basins of a comparatively homogeneous region. For this reason the idea of establishing intensive observation in selected easily accessible basins was widely accepted in the tropics. After several years, however, it was found difficult to utilise data intensively collected outside basin boundaries. One of many possible approaches recommended for the tropics is to use the data for the development of unit hydrograph methods. In addition simulation techniques and a model approach based on intensively collected data have become widely used in recent years.

Still one problem remains: from which region can the data from representative basins be considered as representative and how can one use models fed during the developing stage by data from representative basins, in ungauged or partially gauged regions?

The following approach to the regionalisation and transportation of results, based on system analysis, has been used in practice for various types of projects (Balek [1]):

a) *Identification of requested results*

This originates from discussion between the user of the results and the organisation responsible for their transmission. It is typical that the user is frequently not aware of what can and what cannot be achieved.

b) *Identification of initial information*

This consists of a field survey of the ungauged area and gauged basins, which can possibly be used as an initial source of information. Also, reasonable information can be obtained from the survey of reports, studies, papers etc. relevant to the hydrology and water resources management in the area.

c) *Identification of available data*

The main source of information are data and records published in yearbooks or stored in archives of hydrological survey and water affairs organisations, and temporary records collected by consulting companies, universities etc., as part of various research projects. Meteorological offices produce much needed long-term rainfall data.

d) *Identification of studied area and representative and experimental basins*

This is mainly dependent on the budget and time available. Typically it can consist of:

Identification of the boundaries and their relation to hydrological and hydrogeological divides.

Set-up temporary representative basins within the area, including a meteorological network, stream gauging and rain gauging.

Identification of types of soil within the area, determination of soil moisture characteristics and soil maps. Analysis of the vegetational cover, identification of vegetational types and the establishment of temporary measurements of evapotranspiration.

Drilling of test boreholes, establishment of temporary groundwater level observations and performance of pumping tests.

Discharge measurements of gauged and ungauged streams during the period of low flows, with the aim of determining characteristics of baseflow on all streams and the relationship between gauged and ungauged streams.

Special works such as geophysical, geothermal, water quality and infrared aerial photography, also the dating of water samples.

The time necessary for such a work is, under normal circumstances, no less than three years. The density of the network varies from area to area, depending on the quality of available data from gauged basins. In principle it should be always possible to compare short-term records with long-term ones.

e) *Identification of constraints*

Most frequently, short time, limited budget and limited instrumentation/computer facilities are among typical constraints. Also the inability to fully identify the area can negatively affect the transportation of results.

f) *Selection of working methods*

This type of work consists of:

Unification of the short-term period of observation used for the joint evaluation of data. Most frequently this is determined by the length of the observation by the temporary network. This of course sometimes requires the elimination of part of otherwise reliable records.

Selection of long-term rainfall and/or runoff data from gauged basins, which will be used as a base for the extension of short records.

Analysis of climatic features for both short and long-term periods of observation, for the evaluation of the representativeness of the short-term period of observation. Stage – discharge relationships for discharge evaluation in temporarily observed streams.

Separation of the hydrograph components.

Determination of rainfall-runoff relationships.

Analysis of the groundwater regime and regime of springs in relation to the properties of the aquifer.

Analysis of man's activity.

Allocation of the validity of the above results for subregions in the studied area. This is done in accordance with prevailing groundwater regime, soil type, vegetational cover and morphology.

Selection of an adequate model, which can be applied to the area for analytical and for managerial purposes. Alternatively the application of a model can be rejected.

Water balance calculations for hydrologically significant points such as confluences, gauged cross-sections, lowest points of the basins etc.

Water balance calculations for points significant to the project.

g) *Evaluation of natural end exploitable water resources.*

This also includes long-term prediction of the effects of continuous exploitation, prolonged wet and dry periods etc.

h) *Generalisation of the results outside the limits of the studied area.*

Further generalisation is provided for a comparison of the results with results from other areas and thus for the purpose of extended transportation of results.

From the above techniques an increased data collection obtained through remote sensing methods is foreseen in the tropics. These techniques cover a wide range of applications and they have in common the fact that information on various phenomena is collected by registration and the analysis of radiation emitted, reflected or re-emitted from the objects. The main advantage of remote sensing techniques is that large areas can be covered in a short time and at one go. The disadvantage is that without "ground truth" observations, the remote sensing results are far from complete.

Most popular are the remote sensing data collected via satellites through a multispectral scanner which scans the earth using several wavelengths, and through a set of return vidicon tubes which photograph the earth through filters, one that takes images in green, one in red and one in near infrared light. The preferred wavelength bands are green 500 – 600 nm, red 600 – 700 nm, infrared 700 – 800 nm, infrared 800 – 1000 nm.

In addition to satellites, low and high flying aircraft, model planes and distant photographs from ladders, cranes, etc. have become popular.

The surface of the earth is covered with different bodies; each of which has unique properties for the reflectance of electromagnetic radiation. Water bodies look rather dark, forests appear in different shades of green and red depending on age, transpiration, etc.

For more information a publication of Cracknell [4] is recommended. It also provides numerous references.

Methods in the field of regionalization and transportation of results from representative to ungauged basins are at present far from being uniform and standards for such work vary from region to region. Case studies published outside the tropics can be used for guidance only very carefully and only when all conditions, the density of the network and the aims to be achieved are similar.

Valuable information on the regionalization of data, including Rodier's methods, based on the records of sixty representative basins in west and Central Africa, by ORSTOM, is found in a special guidebook published by Unesco [27].

8.8 REFERENCES

[1] Balek, J., 1980. Regionalisation and transport of results from representative basins to ungauged basins. Int. Symp. on the Inf. of Man and the Hydrol. Regional., Helsinki. General Reports, 117—128.

[2] Burchi, S., 1981. Selected legal and institutional issues in water res. management. Water Qual. Bull., Vol. 6, No. 2.

[3] Cano, G. S., 1981. The recommendations of the United Nations Water Conf. (Mar del Plata, 1977) and related meetings on national water laws. Journal of Hydrology 51, 381—392.

[4] Cracknell, A. P., 1980. Remote sensing in meteorology, oceanography and hydrology. J. Wiley, New York, 542 p.

[5] Ellis, K. V., 1979. Waste water treatment in a developing African state. Wat. Poll. Control,79, No. 3, 409—412.

[6] Frankel, R. J., 1981. Design, construction and operation of a new filter approach for treatment of surface waters in Southeast Asia. Journal of Hydrology 51, 319—328.

[7] Geddes, W. R., 1954. The Land Dyaks of Sarawak. H.M.S.O., London.

[8] Gourou, P., 1959. Los paises tropicales. Universidad Veracruzana, 242 p.

[9] Hauska, R., Karström, V., Kihlblum, V., 1978. Remote sensing potential for hydrology in developing countries. In "Hydrol. in Dev. Countries, Ed. Riise and Skofteland", Nordic IHP Rep. No. 2, 163—175.

[10] Huntington, E., 1915. Civilization and climate. New Haven.

[11] Kashef, A. I., 1981. The Nile-one river and nine countries. Journal of Hydrology 53, 53—71.

[12] Kashef, A. I., 1981. Technical and ecological impacts of the High Aswan Dam. Journal of Hydrology 53, 73—84.

[13] Kirkby, A., 1977. Primitive irrigation. In "Geographical Hydrology, Ed. R. J. Chooley", Methuen, 52—55.

[14] Kovda, K., 1972. Arid lands irrigation and soil fertility. In "Arid Land Irrigation in Dev. Countries", Pergamon Press, Oxford, 211—236.

[15] Mann, H. S., Chatterji, P. C., 1978. Impact of mining operation on the ecosystem in Rajasthan, India. Int. Congress for Energy and Ecosystem, Grand Forks. Tech. Rep. Vol. 2., 615—626.

[16] Manly, K. H., Hafet, M., 1979. The river Nile. Water Qual. Bull., Vol. 4, No. 4., 73—77.

[17] Omari, C. K., 1981. To bring adequate supplies of water to a village. Rolex Awards for Enterprise, Geneva.

[18] Pickford, J. A., 1979. Control of pollution and disease in developing countries. Water Poll. Control, 78, No. 2, 239—253.

[19] Putney, A. D., 1980. Overview of conservation in the Caribbean region. Wildlife Man. Inst. N. Amer. Wildlife and Nat. Res. 54th Conf., Miami Beach, 460—468.

[20] Roberts, T. R., 1981. The ichthyological exploitation of Borneo. Rolex Awards for Enterprise, Geneva.

[21] Rossi, B. A. B., 1981. A complete resource recovery system for solid wastes. Rolex Awards for Enterprise, Geneva.

[22] Tschannerl, G., 1972. Need of hydrol. and hydromet. data for large-scale and long-term investment. Symp. on the role of hydrol. and hydromet. on the econ. dev. of Africa. WMO Rep. No. 301, Geneva, 37—46.

[23] Tschannerl, G., 1958. Notes on temperature -humidity index. U.S.W.B.L.S. 5922, Washington D.C., 4 p.

[24] Tschannerl, G., 1979. Water management and environment in Latin America. UN Econ. Comm. for Latin Amer. Repo., Vol. 12, 359 p.

[25] Verboom, W. C., 1971. Farming and environmental pollution. Univ. of Zambia, Lusaka, 34—35.

[26] Wortington, E. B., 1972. Arid land irrigation in developing countries. Pergamon Press, Oxford.

[27] Anon., 1982. Application of results from experimental and representative basins. Unesco Press, 477 p.

SUBJECT INDEX